Microscopic Imaging Through Turbid Media

BIOLOGICAL AND MEDICAL PHYSICS, BIOMEDICAL ENGINEERING

The fields of biological and medical physics and biomedical engineering are broad, multidisciplinary and dynamic. They lie at the crossroads of frontier research in physics, biology, chemistry, and medicine. The Biological and Medical Physics, Biomedical Engineering Series is intended to be comprehensive, covering a broad range of topics important to the study of the physical, chemical and biological sciences. Its goal is to provide scientists and engineers with textbooks, monographs, and reference works to address the growing need for information.

Books in the series emphasize established and emergent areas of science including molecular, membrane, and mathematical biophysics; photosynthetic energy harvesting and conversion; information processing; physical principles of genetics; sensory communications; automata networks, neural networks, and cellular automata. Equally important will be coverage of applied aspects of biological and medical physics and biomedical engineering such as molecular electronic components and devices, biosensors, medicine, imaging, physical principles of renewable energy production, advanced prostheses, and environmental control and engineering.

More information about this series at http://www.springer.com/series/3740

Min Gu · Xiaosong Gan · Xiaoyuan Deng

Microscopic Imaging Through Turbid Media

Monte Carlo Modeling and Applications

Springer

Min Gu
Centre for Micro-Photonics, Faculty
of Science, Engineering and Technology
Swinburne University of Technology
Hawthorn, VIC
Australia

Xiaosong Gan
Centre for Micro-Photonics, Faculty
of Science, Engineering and Technology
Swinburne University of Technology
Hawthorn, VIC
Australia

Xiaoyuan Deng
Research Resource Centre
South China Normal University
Guangzhou
China

ISSN 1618-7210 ISSN 2197-5647 (electronic)
Biological and Medical Physics, Biomedical Engineering
ISBN 978-3-662-46396-3 ISBN 978-3-662-46397-0 (eBook)
DOI 10.1007/978-3-662-46397-0

Library of Congress Control Number: 2015932069

Springer Heidelberg New York Dordrecht London

Printed on acid-free paper

Springer-Verlag GmbH Berlin Heidelberg is part of Springer Science+Business Media
(www.springer.com)

To our families

Preface

Biophotonics is a new discipline in biophysics and has emerged from research and development at the interface between the photonics and biomedical science and engineering. It covers all the processes initiated by quanta of light and occurring in biological objects as well as the whole complex of optical methods being employed to investigate them. Biophotonics is a photonic solution for biotechnology and medicine and therefore has become an important discipline in life sciences. Imaging through tissue-like turbid media is a fundamental topic in biophotonics. Optical microscopic imaging has played a crucial role in this area. However, due to the multiple scattering effect in a turbid medium, the standard microscopic imaging theory based on diffraction theory is not necessarily applicable in this case. An alternative way for understanding the microscopic imaging performance is based on Monte Carlo simulation that involves Mie scattering.

Research work on Monte Carlo simulation for optical microscopy in tissue media was initiated immediately after the first two authors joined Victoria University in 1995, where they developed the Monte Carlo methodology for optical microscopy. A systemic investigation into various microscopic imaging/gating methods including angle gating, confocal gating, and polarization gating was conducted during that period. As soon as the three authors joined Swinburne University in 2000, the research work on this topic was extended to coherence gating, fluorescence gating, and multi-photon fluorescence gating. The aim of this book is to provide a systematic introduction into these methods. The book can be used by research students, scientists, and engineers who are interested in biophotonics, optical microscopic imaging through tissue, and Monte Carlo simulation.

Although this book was completed by the three authors, many people made important contributions to the topic. We would like to acknowledge the significant contribution by Dr. Steven Schilders who, as a Ph.D. student at Victoria University in Australia, investigated the angle-gating, polarization-gating, and fluorescence-gating methods. Mr. Aernout Kisteman, a visiting research student from the University of Twente in Netherlands, conducted a careful experiment which determined the penetration depth of the single-photon and two-photon fluorescence-gating methods. Dr. Qiang Lu, a visiting scientist from Huazhong University of Science and Technology in China, made an important contribution to the coherence-gating method when he visited Swinburne University of Technology, Australia. We would also like to thank other students and colleagues from Victoria University and Swinburne University of Technology who gave us comments and suggestions on the book. Special acknowledgments are given to South China Normal University, which has supported the third author to complete this book project after 2003. Finally, we would like to acknowledge the important support from the Australian Research Council on this research topic.

Australia Min Gu
Australia Xiaosong Gan
China Xiaoyuan Deng

Contents

Chapter 1
Introduction

Abstract The physical foundation of imaging through tissue is light scattering by small particles because a tissue medium is a diffusing turbid medium that consists of many scatterers such as cells and nuclei. A light beam incident upon a tissue-like turbid medium can be multiply scattered by small particles. As a result, methods for investigating the light-tissue interaction process and the performance of imaging systems such as an optical microscope are different from those based on Fourier optics (Born and Wolf, Principles of optics, 1980; Goodman, Introduction to fourier optics, 1968; Gu, Advanced optical imaging theory, 2000; Wilson and Sheppard, Theory and practice of scanning optical microscopy, 1984; Gu, Principles of three-dimensional imaging in confocal microscopes, 1996). In this introductory chapter, we first describe the physical property of a scattered light beam in Sect. 1.1. In Sect. 1.2, a particular method for investigating light-tissue interaction, called Monte Carlo simulation, is briefly introduced. The main issues related to microscopic imaging through turbid media are summarized in Sect. 1.3. Section 1.4 discusses the two aspects of microscopic imaging through turbid media, the direct and inversed approaches. Finally, the structure of this book is outlined in Sect. 1.5.

1.1 Physical Difference Between Scattered and Unscattered Photons

There has been a substantial increase [1–5] in research into imaging with nonionizing radiation (e.g., laser emission) [6, 7]. Researches are currently trying to develop techniques and theoretical models to help in the imaging of very small tumors embedded in thick layers of human tissue for medical applications (e.g., optical tomography and skin biopsies). Tissue is highly diffusive and therefore acts as a highly scattering turbid medium, which creates problems in detecting the necessary light signal to form an informative image on the scale required. This is due to the nature of the detected illumination photons once they have propagated through

M. Gu et al., *Microscopic Imaging Through Turbid Media*,
Biological and Medical Physics, Biomedical Engineering,
DOI 10.1007/978-3-662-46397-0_1

a highly scattering turbid medium. Consequently, obtaining an image of structures embedded within or behind turbid media has remained to be one of the most challenging problems in the fields of physics, biology, and medical diagnostics.

1.1.1 Classification of Photons

It is well known that the detected illumination photons originating from turbid media consist of unscattered (or ballistic), snake, and multiply scattered components [8]. The unscattered component (Fig. 1.1a) travels in a straight line and traverses the shortest distance through turbid media. Unscattered photons retain the characteristics of the incident light and carry the maximum information about the structures embedded within or behind turbid media. The snake component (Fig. 1.1b) consists of photons that propagate along zigzag paths slightly off the straight-line unscattered path. The snake photons retain significant properties of the incident light and information about structures embedded within or behind turbid media. The multiply scattered component (Fig. 1.1c) consists of photons randomly scattered at various angles in turbid media. Multiply scattered photons travel long distances through turbid media and emerge later in time and in all directions. These photons lose many of their initial physical characteristics and carry little information about structures embedded within or behind turbid media. Multiply scattered photons are the source of image blurring and resolution deterioration that make it difficult to obtain the necessary information needed for high contrast and high resolution imaging. The degradation of the image quality can become so severe in turbid media that the embedded structures are completely hidden from view.

1.1.2 Physical Properties of Photons

There are a few physical effects that need to be considered regarding the photon components mentioned in the previous section. These effects are associated with the spectral, spatial, temporal, and polarization properties of the illumination light propagating through turbid media. The spectral property describes the frequency shift of the illumination photon flux. The deviation (spreading) of the illumination photon flux in space is described by the spatial property of scattered photons. The temporal property describes the time-induced delay (pulse spreading) of the illumination photon flux. The polarization property describes the phase relationship of individual photons in the illumination photon flux. All these effects are mutually connected and are independent of the light source used except for the temporal effect, which cannot be detected with a continuous-wave laser source (i.e., a pulse light source is needed).

The spectral effect is demonstrated in Fig. 1.2a for pulsed illumination. The spectral effect can only be considered if inelastic collisions take place [9]. After the

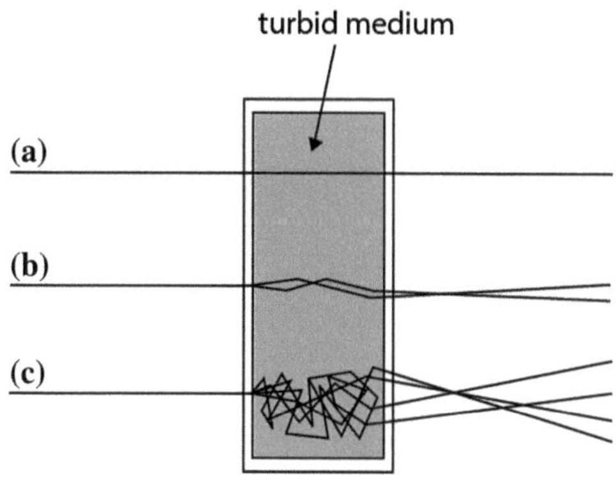

Fig. 1.1 Photon propagation within turbid media. For a clear description, the three components are depicted separately: **a** unscattered photons; **b** snake photons; **c** multiply scattered photons

illumination pulse passes through a turbid medium the snake and multiply scattered photons will have their frequencies shifted from the unscattered component. The magnitude of the spectral shift, $\Delta\omega$, induced depends on the number of scattering events experienced by an individual photon. That is, the larger the number of scattering events experienced the larger the spectral shift incurred. Most of the present research tends to treat collisions as an elastic process, which means that there is no energy transfer from photons to scattering particles, so no spectral change occurs (i.e., $\Delta\omega = 0$). This situation simplifies the problem of modeling, since it ignores scattering phenomena such as Mandel'stam-Brillouin and Raman scattering [9]. Although no scattering event is purely elastic, in most situations this is a good and reasonable approximation to be made since the frequency shift, $\Delta\omega$, is small. Throughout this book, we assume that there is no frequency shift.

The spatial spreading of an incident beam with a Guassian beam profile is illustrated in Fig. 1.2b. The collected unscattered portion of the illumination beam is illustrated with dashed lines. Snake and multiply scattered photons cause the diffraction pattern generated by the unscattered photons to broaden, since they travel along paths that are different from the unscattered straight through path (see Fig. 1.1). The amount of spatial broadening of the illumination beam is determined by the characteristics of the turbid media at the wavelength of illumination. The statistically different propagation paths of unscattered and scattered photons are the bases of the angle-gating principle discussed in Chap. 5.

The temporal properties can only be considered when a pulsed illumination beam is considered (Fig. 1.2c). An incident pulse can broaden (in the time domain) when the pulse undergoes its transition through turbid media. The unscattered photons are the first to arrive, followed closely by the snake photons, and at a later time the multiply scattered photons arrive. The turbid media characteristics at the wavelength of illumination determine the amount of delay (temporal pulse broadening) induced between the unscattered, snake, and scattered photons.

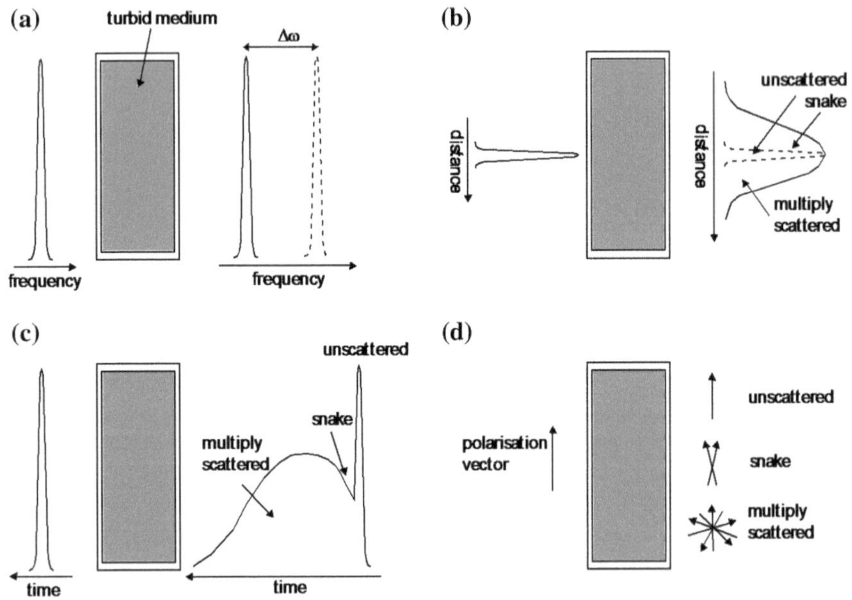

Fig. 1.2 Schematic demonstration of four physical effects of photons propagating through turbid media: **a** spectral shift; **b** spatial broadening; **c** temporal broadening; **d** polarization degradation

Figure 1.2d demonstrates the change in the polarization vector of an illumination beam propagating through turbid media. The state of polarization for the illumination beam (for example, defined by the vertical polarization vector in Fig. 1.2d) is maintained with unscattered photons. Snake photons lose some degree of the illumination polarization state, and the multiply scattered photons suffer substantial depolarization. That is, the orientation of the polarization vector is partially or completely random. The magnitude of the direction change in the polarization vector is determined by the characteristics of turbid media at the wavelength of illumination.

As summarized in Table 1.1, these four effects (spectral, spatial, temporal, and polarization) provide direct distinguishable differences between the unscattered, snake, and scattered components, which can be taken advantage of when a particular detection scheme is used. One can assume that unscattered photons and snake photons are those necessary to create an informative high resolution and high contrast image. Thus, methods for detecting only unscattered and snake photons (the coherent component of the illumination beam) that carry more information about the embedded object, while the multiply scattered photons (the incoherent component of the illumination beam) are suppressed, have been used in imaging through turbid media. However, it should be noted that images may be reconstructed from the multiply scattered photons if the phase and amplitude of the scattered photons are known at many points in space. However, the reconstruction

Table 1.1 Physical difference between unscattered and scattered photons after propagating through a turbid medium

Unscattered photons provide	Scattered photons provide
• Early arrival time	• Late arrival time
• Same coherence	• Low coherence
• Same polarization	• Depolarized
• Same direction	• Different direction

of images from multiply scattered photons, known as the inverse scattering problem [10], remains a difficult experimental and theoretical problem to be overcome in imaging through turbid media.

1.2 Microscopic Imaging Through Tissue-Like Media

Imaging an object embedded in a turbid medium has attracted substantial interest since it is potentially related to applications in early cancer detection. Research work in this field can be classified into two categories: transillumination imaging, in which case a parallel beam probe is used [11–20] and microscopic imaging, in which a microscopic objective is used for illumination as shown in Fig. 1.3 [21–58].

In an optical microscope, an object embedded in a turbid medium is illuminated by an objective lens of a range of the illumination angle. The optical signal in each direction comprises two parts; one part is scattered by the embedded object and the other is scattered by the scattering medium surrounding the object. The former is a wanted signal carrying information about the object and forms an image by an imaging objective, while the latter is unwanted as it mainly contributes to the background of an image. As a result of using an illumination objective, the two parts of the signal superpose each other, which degrades the image quality. A number of approaches have been proposed to obtain useful images through significant depths of a turbid medium. A gating method means the suppression of the unwanted scattered signal. The gating methods currently available to selectively suppress the scattered photons based on the properties in Table 1.1 are time-gating [8], which relies on the utilization of an ultrashort pulsed beam, coherence-gating [26], which relies on the degree of coherence of photons, polarization-gating [17], which relies on the polarization-state of photons, and angle-gating [33–36], which relies on the path deviation of the scattered photons. Although all of these gating mechanisms can be employed in any imaging system, the efficiency of these methods depends on a particular imaging system. Transillumination imaging systems which use a parallel beam probe can give images of millimeter resolution [6, 7]. To obtain an image of micrometer resolution, a microscope objective is necessary. In this case, time-gating may become less efficient due to the large range of illumination angles. However, angle-gating, polarization-gating, coherence-gating, and fluorescence-gating are important in microscopic imaging. In addition,

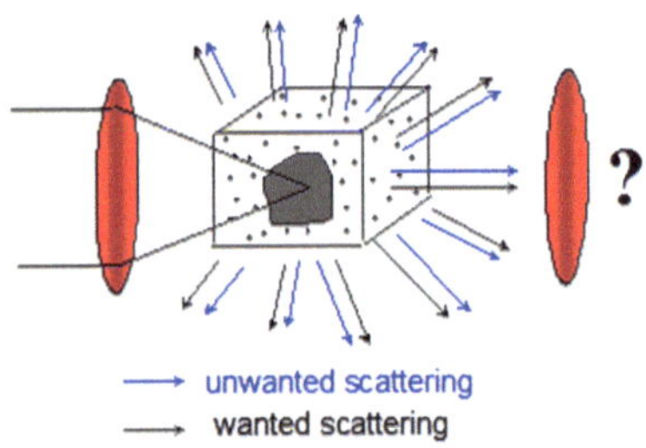

Fig. 1.3 Microscopic imaging through a turbid medium in which an object is embedded

the use of an objective leads to a focal region of an intensity that is high enough to produce nonlinear excitation such as two-photon excitation [27]. Because the strength of the nonlinear signal is mainly determined by the ballistic photons, any nonlinear excitation process under a microscope can be used to suppress scattered photons, which results in a unique nonlinear-gating mechanism in microscope imaging through turbid media [47].

However, using an objective lens in a microscopic imaging system raises the question of which numerical aperture of an objective is suitable for imaging. According to the imaging theory based on Born's approximation, which ignores the multiple scattering in a turbid medium, a high numerical aperture objective lens can provide high diffraction-limited resolution [1–5]. Born's approximation is applicable to the case in which the optical thickness n, defined as the thickness of a turbid medium divided by the scattering mean free path length, is less than one. On the other hand, a low numerical aperture objective can suppress scattered photons that statistically travel at high angles. Both arguments are based on the assumption that ballistic light is dominant in forming an image. When a turbid medium is thick, e.g., when $n > 10$, the strength of the unscattered light/photons may be too weak to be detected, particularly in the presence of detector noise. In this situation scattered light has to be included in constructing an image. An important question raised here is what role the scattered light plays in constructing an image. To answer this question, the relationship of scattered photons to image resolution should be investigated. In other words, suitable numerical aperture of the objective lens to be used in a particular microscopic imaging system for high-quality imaging should be investigated.

Turbid media, typically biological tissue, always exhibits complex characteristics as it has complex structures and is composed of various components. Usually, it shows a multiple-layer structure rather than a single-layer structure and consists of multiple sizes of scatterers rather than a single size. Furthermore, an inhomogeneous feature may exist because of the aggregation effect of the scatterers. Such structural, size, or aggregation features from turbid media will greatly influence images under a microscope.

1.3 Monte Carlo Simulation

Microscopic imaging study through tissue-like media is an important and significant topic since it is potentially related to the use of microscopy in noninvasive imaging and spectroscopy of biological tissues, and thus the potential applications in early medical diagnoses of tumors [59]. However, the analysis of microscopic imaging in a highly scattering turbid medium is complicated by the basic incompatibility between the techniques developed for modeling diffraction-limited optical systems [1–5] and those developed for modeling light propagation in multiply scattering media [9, 60]. For a medium in which the concentration of the scatterers is high enough that multiple scattering cannot be neglected, the diffraction theory fails. The single-scattering theory based on the Born approximation is not appropriate for describing light transport in biological soft tissues thicker than a few tens of micrometers due to the strong scattering in the wavelength range of 0.6–1.0 μm [61]. Some already obtained analytical expressions based on the radiative transfer theory describing the effects of multiple scattering on the performance of imaging systems are either based on the small-angle approximation [62], which is applicable only to a medium containing a relatively low concentration of particles with size much larger than a wavelength, or on the evaluation of a multidimensional integral that is practical to evaluate only for low orders of scattering [63].

Biological tissue is a highly optical scattering material; it contains dense concentrations of anisotropic scatterers and the inhomogeneous cellular structures with scatterers are usually of optical wavelength order. Because of the optical dense feature in biological tissue, the diffuse approximation of the radiative transfer theory can be applied [64]. Since this method deals with mainly forward scattered photons, it is hardly applicable to a microscopic imaging system in which photons scattered into a large angle affect the imaging performance significantly.

The Monte Carlo method is easy to apply and flexible to handle complex geometries and inhomogeneity. By tracing the behavior of various types of photons, ballistic, snake, multiply forward-scattered/diffuse photons, it is helpful to gain an understanding of the underlying physics in photon migration through a tissue-like turbid medium and imaging resolution in an optical system. Therefore, in the recent three decades, Monte Carlo simulation has been given considerable attention and widely applied to the tissue optics [65]. In the study of microscopic imaging of an object deeply embedded in a tissue-like sample, the Monte Carlo method is one that can be successfully and flexibly combined with microscopic systems for investigating imaging performance and image reconstruction [23, 29, 30, 33, 34, 37, 41, 42, 46–48, 50–58, 65].

The Monte Carlo method can be conceptually understood to be a stochastic technique, which means it is based on the use of random numbers and probability statistics to investigate problems, although an accurate, complete, and concise definition to characterize it is difficult to construct [66, 67]. A definition given by Lux et al. [68] may give us a mathematical understanding: in all the applications of the Monte Carlo method, a stochastic model is constructed in which the expected value of a certain random variable (or of a combination of several variables) is equivalent to the value of a physical quantity to be determined. This expected value is then estimated by the average of several independent samples representing the random variable introduced above. For the construction of a series of independent samples, random numbers following the distributions of the variable to be estimated are used.

In the case of Monte Carlo simulation of photon propagation through a scattering medium, individual photons are supposed to experience events of scattering and absorption according to the local values of the optical properties of the scattering medium. The individual photon paths can be simulated by considering the probability distribution of two random variables, the step length and the scattering angle. The probability distributions of the random variables of the step length and scattering angle, respectively, describe the step size a photon may take between two successive photon-medium interaction sites and the angles of deflection a photon may experience when a scattering event occurs. These two probability functions can be derived from Beer's law and the Mie scattering theory [9, 60]. The Monte Carlo simulation then is performed by tracing the random walks that a large number of photons make based on the statistical sampling from the probability distributions in each scattering event.

1.4 Direct and Inverse Approaches

Imaging modeling usually includes two aspects, direct and inverse approaches. In the former approach, one simulates the final image if the object condition and the property of an imaging system are given. In the latter approach, one reconstructs the object function if an image is given. Given that no imaging system is perfect, any measured image shows less detail than the original object. Using an inverse approach, one can enhance the image quality such as resolution and contrast. To this end, it is necessary to have a defined mathematical relation between an object and its image. For example, the imaging theory based on light diffraction shows that a convolution relation holds under the paraxial approximation [1–5].

However, as discussed in the previous section, due to the complicate nature caused by multiple scattering in a turbid medium, conventional image modeling methods based on the diffraction theory [1–5] are not applicable. The Monte Carlo simulation method involving the Mie theory is one of the useful tools for imaging modeling in a turbid medium, because it provides insight into light interaction with scattering particles. In this aspect, the Monte Carlo method is used as a direct approach. However, this type of simulation may not be applicable in image reconstruction because there is no defined mathematical relation between an object and its image. Another problem associated with image modeling based on the Monte Carlo method is time-consuming because a large number of incident photons are required to ensure a required accuracy. The requiring of computational time increases significantly when images of complicated objects embedded in a turbid medium are modeled.

To address these problems, the concept of the effective point spread function (EPSF) has been introduced in transillumination imaging through turbid media. Recently, this concept has been generated to Monte Carlo simulation in a turbid medium under an optical microscope [42]. It should be pointed out that defining a point spread function (PSF) in a turbid medium is not straightforward. If there is no turbid medium, a PSF is the image of a point object and is a measure of image blurring through an imaging system. It determines resolution of an imaging system and therefore is independent of the property of an object. As a result, under Born's approximation [1, 2], the image of an object can be obtained from the convolution of a PSF with an object function. If an object is embedded in a turbid medium, it is difficult to use the concept of a point object because of the existence of scattering particles which are in the range of 0.1–1 μm in diameter. Further, a PSF that includes only the property of a microscope may not be adequate because the multiple scattering effect may severely distort the image of an embedded object. Therefore, the property of a microscope as well as the property of scattering particles should be included in an effective PSF for a microscope.

Such an EPSF reflects not only the property of a microscopic imaging system but also the scattering property of a turbid medium. The parameters determining an imaging system include the numerical aperture (NA) of illumination and detection objectives and the size of confocal pinhole v_d. The parameters that describe a turbid medium are the scattering mean free path length l, the anisotropy value g,, and the optical thickness n which is the sample thickness divided by the scattering mean free path length. Further, such an EPSF should satisfy a convolution operation, i.e., the condition that the image intensity $I(x, y)$ of a thin object can be modeled by the convolution of an object function $O(x, y)$ and the EPSF $h(x, y)$:

$$I(x,y) = \iint_{-\infty}^{\infty} h(x,y)O(x - x', y - y')\mathrm{d}x'\mathrm{d}y', \tag{1.1}$$

where $h(x, y)$ is the EPSF in the focal plane. With the help of the convolution relation, the computational time for image modeling can be dramatically reduced.

Equation (1.1) also provides a tool to perform image reconstruction. As a result, the efficiency of each gating method can be well characterized because gating methods can play an important role in microscopic imaging through turbid media. On the other hand, it has also been demonstrated that signal strength can be insufficient if a significant amount of scattered photons is removed. It is worthwhile to discuss the role of scattered photons. Are multiple scattered photons merely noise and do they make no positive contribution in building an image? The statistical analysis of scattered photon distributions shows that scattered photons still carry information about embedded objects [37, 69]. However, they are always treated as noise when high resolution is pursued. In a thick turbid medium, because of nearly non-existence of ballistic or least scattered photons, multiple scattered photons have to be taken into account in building up an image. This inevitably degrades the image resolution. In this circumstance, the inverse approach (image reconstruction) is regarded as a solution to the problem [52, 53].

1.5 Overview of the Book

This book is organized to describe the Monte Carlo simulation method in microscopic imaging through turbid media. The book consists of nine chapters including this introductory chapter. Chapters 2–9 include three main topics, which are the fundamentals of the Monte Carlo simulation method for an optical microscope in turbid media, the direct approach of imaging modeling, and the inversed approach of imaging modeling. The following brief outline provides an overview of the three topics.

The first topic is the fundamental of the Monte Carlo simulation method for an optical microscopy system in a turbid medium and is covered by Chaps. 2–4. Chapter 2 describes the basic physical concepts and theory of scattering media. The concept of Rayleigh and Mie scattering models and their effect on tissue-like turbid media are discussed. In particular, the Mie scattering theory for a spherical scattering particle and the effective Mie scattering theory for a spherical scattering aggregate are introduced. Chapter 3 provides a detailed description of the Monte Carlo simulation method for an optical microscope. After the general description of the Monte Carlo simulation method, a number of specific effects in an optical microscope are discussed. These effects are divided into two aspects. The first aspect is related to the property of a turbid medium, including the effect of a boundary, a multi-sized scatterer, and scatter aggregation. The second aspect deals with the various treatments of the physical property of an illumination beam in a microscope. These treatments cover polarization, pulsed illumination, coherence, diffraction, and nonlinear excitation. Based on the Monte Carlo simulation in an optical microscope, Chap. 4 presents the concept of the EPSF and the method for deriving the EPSF for a microscope from Monte Carlo simulation.

Chapters 5–8 are dedicated to the direct approaches based on the Monte Carlo simulation method introduced in Chaps. 3 and 4. The focus of this part is to understand four optical gating mechanisms used in microscopic imaging through turbid media, angle-gating, polarization-gating, and fluorescence-gating and nonlinear-gating. The performance of the four gating methods is characterized by two physical parameters, resolution and signal level. In particular, the angle-gating mechanism described in Chap. 5 is implemented by two methods, annular objectives and confocal pinhole. Chapter 6 presents two polarization-gating methods, conventional polarization-gating and differential polarization-gating. A high numerical aperture objective is usually needed to perform fluorescence microscopy in tissue sample. Such a microscopy method exhibits a property of suppressing scattered photons. The effects of numerical aperture, confocal pinhole, scatterer size, layered samples, and scatterer aggregation are investigated according to resolution and signal level derived using the Monte Carlo simulation method. The gating mechanism provided by multi-photon fluorescence microscopy is based on the use of ballistic photons. The performance of such a nonlinear-gating method, simulated by the Monte Carlo method, is presented in Chap. 8.

The inversed approach to image reconstruction is introduced in Chap. 9. Images of single point and two-point objects, and ring structures are used to demonstrate the efficiency of the reconstruction method based on the expectation maximization method. The effect of the noise level on the quality of a restored image is also discussed.

References

1. M. Born, E. Wolf, *Principles of Optics* (Pergamon, New York, 1980)
2. J.W. Goodman, *Introduction to Fourier Optics* (McGraw-Hill, New York, 1968)
3. M. Gu, *Advanced Optical Imaging Theory* (Springer, Heidelberg, 2000)
4. T. Wilson, C.J.R. Sheppard, *Theory and Practice of Scanning Optical Microscopy* (Academic, London, 1984)
5. M. Gu, *Principles of Three-dimensional Imaging in Confocal Microscopes* (World Scientific, Singapore, 1996)
6. B. Tromberg, A. Yodh, E. Sevick, D. Pine, Diffusing photons in turbid media: introduction to the feature. Appl. Opt. **36**, 9 (1997)
7. A. Yodh, B. Tromberg, E. Sevick-Muraca, D. Pine, Introduction to the special issue on diffusing photons in turbid media. J. Opt. Soc. Am. A **14**, 136 (1997)
8. E.B. de Haller, Time-resolved transillumination and optical tomography. J. Biomed. Opt. **1**, 7 (1996)
9. C.F. Bohern, D.R. Huffman, *Absorption and Scattering of Light by Small Particles* (Wiley, New York, 1983)
10. H.P. Baltes, *Inverse Source Problems in Optics* (Springer, Berlin, 1978)
11. D.G. Papaioannou, G.W. t Hooft, J.J.M. Baselman, M.J.C. van Gemert, Image quality in time-resolved transillumination of highly scattering medium. Appl. Opt. **34**, 6144 (1995)
12. J.G. Fujimoto, S. De Silvestri, E.P. Ippen, C.A. Puliafito, R. Margolis, A. Oseroff, Femtosecond optical ranging in biological systems. Opt. Lett. **3**, 150 (1986)
13. S.L. Jacques, Time resolved propagation of ultrashort laser pulses within turbid tissue. Appl. Opt. **28**, 2223 (1989)
14. D. Huang, E.A. Swanson, C.P. Lin, J.S. Schuman, W.G. Stinson, W. Chang, M.R. Hee, T. Flott, K. Gregory, C.A. Pulianfito, J.G. Fujimoto, Optical coherent tomography. Science **254**, 1178 (1991)
15. M.J. Yadlowsky, J.M. Schmitt, R.F. Bonner, Multiple scattering in optical coherence microscopy. Appl. Opt. **34**, 5699 (1995)
16. S.G. Demos, R.R. Alfano, Optical polarization imaging. Appl. Opt. **36**, 150 (1997)
17. S.P. Morgan, M.P. Khong, M.G. Somekh, Effects of polarization state and scatterer concentration on optical imaging through scattering media. Appl. Opt. **36**, 1560 (1997)
18. K.M. Yoo, R.R. Alfano, Time-resolved coherent and incoherent components of forward light scattering in random media. Opt. Lett. **15**, 320 (1990)
19. S. Anderson-Engels, R. Berg, O. Jarlmann, S. Svanberg, Time-resolved transillumination for medical diagnostics. Opt. Lett. **15**, 1179 (1990)
20. Q.Z. Wang, X. Liang, L. Wang, P.P. Ho, R.R. Alfano, Fourier spatial filter acts as a temporal gating for light propagating through a turbid medium. Opt. Lett. **20**, 1498 (1995)
21. M. Minsky, Microscopy apparatus, US patent 3012467, 1961

22. M. Kempe, W. Rudolph, E. Welsch, Comparative study of confocal and heterodyne microscopy for imaging through scattering media. J. Opt. Soc. Am. A **13**, 46 (1996)
23. J.M. Schmitt, A. Knuttel, M. Yadlowsky, Confocal microscopy in turbid media. J. Opt. Soc. Am. A **11**, 2226 (1994)
24. M. Gu, T. Tannous, C.J.R. Sheppard, Effect of an annular pupil on confocal imaging through highly scattering media. Opt. Lett. **21**, 312 (1996)
25. M. Kempe, A.Z. Genack, W. Rudolph, P. Dorn, Ballistic and diffuse light detection in confocal and heterodyne imaging systems. J. Opt. Soc. Am. A **14**, 216 (1997)
26. J.F. de Boer, T.E. Miller, M.J.C. van Gemert, J.S. Nelson, Two-dimensional birefringence imaging in biological tissue by polarisation-sensitive optical coherence tomography. Opt. Lett. **22**, 934 (1997)
27. W.J. Denk, J.H. Strickler, W.W. Webb, Two-photon laser scanning fluorescence microscopy. Science **248**, 73 (1990)
28. B.R. Masters, P.T.C. So, E. Gratton, Multiphoton excitation fluorescence microscopy and spectroscopy of in vivo human skin. Biophys. J. **72**, 2405 (1997)
29. V. Daria, C.M. Blanca, O. Nakamura, S. Kawata, C. Saloma, Image contrast enhancement for two-photon fluorescence microscopy in a turbid medium. Appl. Opt. **37**, 7960 (1998)
30. C.M. Blanca, C. Saloma, Monte Carlo analysis of two-photon fluorescence imaging through a scattering medium. Appl. Opt. **37**, 8092 (1998)
31. Y. Guo, Q.Z. Wang, N. Zhadin, F. Liu, S. Demos, D. Calistru, A. Tirksliunas, A. Katz, Y. Budansky, P.P. Ho, R.R. Alfano, Two-photon excitation of fluorescence from chicken tissue. Appl. Opt. **36**, 968 (1997)
32. A.K. Dunn, V.P. Wallace, M. Coleno, M.W. Berns, B.J. Tromberg, Influence of optical properties on two-photon fluorescence imaging in turbid samples. App. Opt. **39**, 1194 (2000)
33. X. Gan, S. Schilders, M. Gu, Combination of annular aperture and polarisation-gating methods for efficient microscopic imaging through a turbid medium: theoretical Analysis. Microsc. Microanal. **3**, 495 (1997)
34. X. Gan, M. Gu, Temporal, angular and spatial distribution of photon migration through a highly scattering medium. Optik **108**, 129 (1998)
35. S.P. Schilders, X.S. Gan, M. Gu, Efficient suppression of diffusing photons using polarising annular objectives for microscopic imaging through turbid media. Bioimaging **6**, 92 (1998)
36. S.P. Schilders, X.S. Gan, M. Gu, Microscopic imaging through a turbid medium using annular objectives for angle-gating. Appl. Opt. **37**, 5320 (1998)
37. X. Gan, S. Schilders, M. Gu, Image formation in a turbid medium under a microscope. J. Opt. Soc. Am. A **15**, 2052 (1998)
38. S.P. Schilders, X.S. Gan, M. Gu, Resolution improvement in microscopic imaging through turbid media based on differential polarisation. Appl. Opt. **37**, 4300 (1998)
39. S.P. Schilders, X. Gan, M. Gu, Image enhancement in a reflection optical microscope by suppression of diffusing photons using polarising annular objectives. Microsc. Microanal. **4**, 415 (1998)
40. S.P. Schilders, X. Gan, M. Gu, Effect of scatterer size on microscopic imaging in turbid media based on differential polarization gating. Opt. Commun. **157**, 238 (1998)
41. X. Gan, S. Schilders, M. Gu, Image enhancement through turbid media under a microscope using polarization gating methods. J. Opt. Soc. Am. A **16**, 2177 (1999)
42. X. Gan, M. Gu, Effective point spread function for fast image modelling and processing in microscopic imaging through turbid media. Opt. Lett. **24**, 741 (1999)
43. S. Schilders, M. Gu, Three-dimensional autofluorescence spectroscopy of rat skeletal muscle tissue under two-photon excitation. Appl. Opt. **38**, 720 (1999)
44. S. Schilders, M. Gu, Limiting factors on image quality in imaging through turbid media under single-photon and two-photon excitation. Microsc. Microanal. **6**, 156 (2000)
45. M. Gu, S. Schilders, X. Gan, Two-photon fluorescence imaging of microspheres embedded in turbid media. J. Mod. Opt. **47**, 959 (2000)
46. X. Gan, M. Gu, Spatial distribution of single-photon and two-photon fluorescence light in scattering media: Monte Carlo simulation. Appl. Opt. **39**, 1580 (2000)

47. X. Gan, M. Gu, Fluorescence microscopic imaging through tissue-like turbid media. J. Appl. Phys. **87**, 3214 (2000)
48. M. Gu, X. Gan, A. Kisteman, M. Xu, Comparison of penetration depth between single-photon excitation and two-photon excitation in imaging through turbid tissue media. Appl. Phys. Lett. **77**, 1551 (2000)
49. M. Xu, E.D. Williams, E.W. Thompson, M. Gu, Effect of handling and fixation processes on fluorescence spectroscopy of rat skeletal muscles under two-photon excitation. Appl. Opt. **39**, 6312 (2000)
50. X. Deng, E.D. Williams, E.W. Thompson, X. Gan, M. Gu, Second-harmonic generation from biological tissues: effect of excitation wavelength. Scanning **24**, 175 (2002)
51. X. Deng, X. Gan, M. Gu, Multi-photon fluorescence microscopic imaging through double-layered turbid tissue media. J. Appl. Phys. **91**, 4659 (2002)
52. X. Gan, M. Gu, Microscopic image reconstruction through tissue-like turbid media. Opt. Commun. **207**, 149 (2002)
53. X. Gan, M. Gu, Image reconstruction through turbid media under a transmission-model microscope. J. Biomed. Opt. **7**, 372 (2002)
54. X. Deng, X. Gan, M. Gu, Monte-Carlo simulation of multi-photon fluorescence microscopy imaging through inhomogeneous tissue-like turbid media. J. Biomed. Opt. **8**, 400 (2003)
55. X. Deng, M. Gu, Penetration depth of single-, two- and three-photon fluorescence microscopic imaging through human cortex structures: Monte-Carlo simulation. Appl. Opt. **42**, 3321 (2003)
56. X. Deng, X. Gan, M. Gu, Effective Mie scattering of a spherical aggregate and its application in turbid media. Appl. Opt. **43**, 2925 (2004)
57. Q. Lu, X. Gan, M. Gu, Q. Luo, Monte Carlo modeling of optical coherence tomography imaging through turbid media. Appl. Opt. **43**, 1628 (2004)
58. M. Gu, X. Gan, in *Image Reconstruction Through a Tissue-Like Turbid Medium Under a Fluorescence Microscope, Optics Within Life Sciences*, ed. by C.J.R. Sheppard (Springer, Heidelberg, 2000)
59. R. Alfano, S. Demos, Advances in optical imaging of biomedical media. Ann. NY Acad. Sci. **820**, 248 (1997)
60. A. Ishimaru, *Wave Propagation and Scattering in Random Media* (Academic, New York, 1978)
61. W.F. Cheong, S.A. Prahl, A.J. Walsh, A review of the optical properties of biological tissues. IEEE J. Quantum Electron. **26**, 2166 (1990)
62. Y. Kuga, A. Ishimaru, Modulation transfer function of layered inhomogeneous random media using the small-angle approximation. Appl. Opt. **25**, 4328 (1986)
63. P. Bruscaglioni, G. Milloni, G. Zazzanti, On the contribution of multiple scattering to lidar returns from homogeneous fogs and its dependence on the receiver angular aperture. Opt. Acta **27**, 1229 (1980)
64. M.S. Patterson, B. Chance, B.C. Wilson, Time-resolved reflectance and transmittance for the non-invasive measurement of tissue optical properties. Appl. Opt. **28**, 2331 (1989)
65. J.M. Schmitt, K. Ben-Letaief, Efficient Monte Carlo simulation of confocal microscopy in biological tissue. J. Opt. Soc. Am. A **13**, 952 (1996)
66. V. Tuchin, *Handbook of Optical Biomedical Diagnostics* (SPIE Press, Bellingham, 2002)
67. J. Spanier, E.M. Gelbard, *Monte Carlo Principles and Neutron Transport Problems* (Addison-Wesley, Reading, 1969)
68. I. Lux, L. Koblinger, *Monte Carlo Particle Transport Methods: Neutron and Photon Calculations* (CRC Press, Boca Raton, 1991)
69. A. Dunn, C. DiMarzio, Efficient computation of time-resolved transfer functions for imaging in turbid media. J. Opt. Soc. Am. A **13**, 65 (1996)

Chapter 2
Scattering of Light by Small Particles

Abstract Scattering properties such as the scattered direction and the strength of an incident electromagnetic wave can be derived from the Mie scattering theory and form a physical basis of the Monte Carlo simulation method used in this book. After the brief explanation on the physical difference between Rayleigh and Mie scattering, presented in Sect. 2.1, the Mie scattering theory and the Stokes vector method that are needed to derive the polarization nature of a scattered field are described in Sects. 2.2 and 2.3, respectively.

2.1 Rayleigh Scattering and Mie Scattering

Optical scattering by a small particle can be divided into two classes according to the size of scatterers. Rayleigh scattering is referred to a scattering process when the size of a scatterer is much smaller than the illumination wavelength. The strength of the scattered field is independent of the scattered direction, which means isotropic scattering. The strength of Rayleigh scattering is inversely proportional to the fourth power of the illumination wavelength. The second class of optical scattering is called Mie scattering in which case the size of a scatterer is comparable to the illumination wavelength. As a result, Mie scattering is an anisotropic scattering. The strength of Mie scattering is related to the scatterer size and the illumination wavelength (Fig. 2.1).

Biological tissue is usually composed of small scatterers such as bacteria, viruses, malignant cells and so on. The size of these scatterers varies from 0.1 μm to a few micrometers [1, 2]. Therefore, the dominant scattering effect caused by these scatterers is Mie scattering rather than Rayleigh scattering. The Mie scattering theory [3–5] provides an important physical foundation of the Monte Carlo simulation for optical microscopic imaging through a tissue-like turbid medium consisting of scattering particles such as cells and nuclei. However, these small scattering particles are sometimes aggregated because of their biological and chemical functions [6, 7]. A fractal aggregate is a medium with an ensemble of

M. Gu et al., *Microscopic Imaging Through Turbid Media*,
Biological and Medical Physics, Biomedical Engineering,
DOI 10.1007/978-3-662-46397-0_2

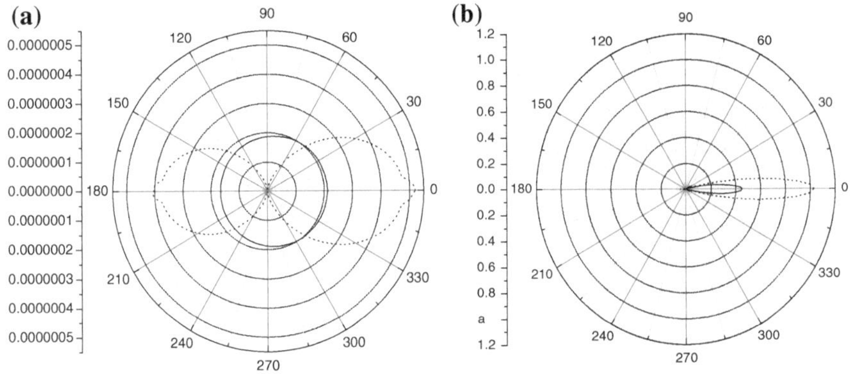

Fig. 2.1 Scattered intensity distribution around a dielectric particle (n = 1.51) of different radii: **a** $a = 0.05$ μm, and **b** $a = 1$ μm. The *solid* and *dotted curves* correspond to the S and P polarization states of the illumination wave (λ = 632.8 nm), respectively

small scattering particles which, due to the physical mechanisms of aggregation, lead to a fractal structure, i.e., a scale-invariant structure [8]. Recently, an effective Mie scattering theory has been developed to derive the scattering property of a single spherical aggregate [9]. In this method, scattering properties of a single spherical aggregate consisting of Mie scatterers is derived from the Monte Carlo simulation. Therefore, such a method will be introduced in Chap. 3 after the introduction of the Monte Carlo simulation.

2.2 Mie Scattering Theory

There are many methods to describe the interaction of light and matter. Mie scattering theory describes scattering of a plane wave by a small particle with an insight into the collision process between photons and scattering particles. The scattering property of a scatterer forms a physical base of a scattering medium or a turbid medium which consists of a large number of scattering particles randomly dispersed.

As shown in Fig. 2.2, when a beam of light with specified characteristics illuminates a given small particle, the amount of the light scattered by the particle along with its angular distribution depends directly on the characteristics of the particle, including its shape, size, and the materials of which it is composed. The physics and mathematics for the interaction of an electromagnetic wave with a sphere is complicated and cumbersome and has been well documented in a number of classic textbooks [3–5]. Therefore, only the Mie theory formulations of scattering parameters including the scattering coefficients, scattering cress-section, anisotropy value, scattering efficiency, and scattering mean free path length, which are needed to calculate the scattering parameters, are presented in this section.

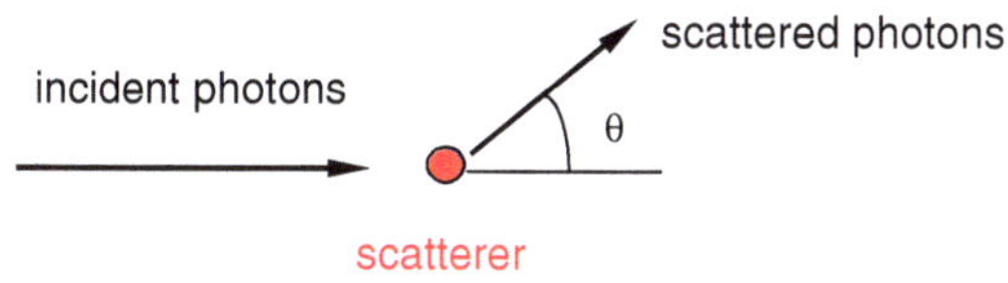

Fig. 2.2 An incident plane wave is scattered by a small particle. The scattered photons propagate at angle θ with respect to the incident direction

2.2.1 Scattering Coefficients of a Spherical Particle

For a spherical particle of a given particle size, the size parameter, x, and the relative refractive index, m, are given by [3]

$$x = k_2 a = \frac{2\pi n_2 a}{\lambda} = 2\pi n_2 A, \tag{2.1}$$

$$m = \frac{k_2}{k_1} = \frac{n_2}{n_1}, \tag{2.2}$$

where k is the wave vector, a is the radius of the spherical particle, n is the refractive index and $A = a/\lambda$ is the relative particle size. Here the subscripts 1 and 2 represent the surrounding medium and the scattering particle, respectively.

The scattering coefficients given by the Mie theory [3], a_i and b_i, which represent the weighting parameters for the electromagnetic normal modes scattered by a spherical particle defined by (2.1), are given by

$$a_i = \frac{\mu_1 m^2 j_i(mx)[xj_i(x)]' - \mu_2 j_i(x)[mxj_i(mx)]'}{\mu_1 m^2 j_i(mx)\left[xh_i^{(1)}(x)\right]' - \mu_2 h_i^{(1)}(x)[mxj_i(mx)]'}, \tag{2.3}$$

and

$$b_i = \frac{\mu_2 j_i(mx)[xj_i(x)]' - \mu_1 j_i(x)[mxj_i(mx)]'}{\mu_2 j_i(mx)\left[xh_i^{(1)}(x)\right]' - \mu_1 h_i^{(1)}(x)[mxj_i(mx)]'}, \tag{2.4}$$

respectively, where the prime indicates differentiation with respect to the argument in the parenthesis. Here $j_i(mx)$ and $h_i^{(1)}(x)$ are the spherical Bessel functions of the first and third kind, respectively, given by

$$j_i(mx) = \sqrt{\frac{\pi}{2mx}} J_{i+1/2}(mx), \tag{2.5}$$

and

$$h_i^{(1)}(x) = j_i(x) + iy_i(x), \tag{2.6}$$

where $y_i(x)$ is a spherical Bessel function of the second kind

$$y_i(x) = \sqrt{\frac{\pi}{2x}} Y_{i+1/2}(x). \tag{2.7}$$

Here i denotes the i-th mode of an electromagnetic wave. In the above definition, $J_{j+1/2}(x)$ and $Y_{j+1/2}(x)$ are Bessel functions of the first and second kind, respectively. The detail of Bessel functions can be found from the standard textbook [10].

2.2.2 Scattering Cross-Section and Anisotropy Value

The scattering cross-section, σ_s, represents the cross-sectional area within which a scattering particle can interact with the illumination light. The scattering cross-section, σ_s, for a spherical particle, defined by (2.1), is given by [3]

$$\sigma_s = \frac{2\pi}{k^2} \sum_{i=1}^{\infty} (2i+1)\left(|a_i|^2 + |b_i|^2\right). \tag{2.8}$$

The anisotropy value (or the asymmetry parameter), g $(=\langle\cos\theta\rangle)$, which defines the averaged directional change of the scattered light, is given by [3]

$$gQ_s = \frac{4}{x^2}\left[\sum_{i=1}^{\infty} \frac{i(i+2)}{i+1} Re\left\{a_i a_{i+1}^* + b_i b_{i+1}^*\right\} + \sum_{i=1}^{\infty} \frac{2i+1}{i(i+1)} Re\left\{a_i b_i^*\right\}\right], \tag{2.9}$$

where Q_s is the scattering efficiency defined as

$$Q_s = \frac{\sigma_s}{\sigma_g} \tag{2.10}$$

and σ_g is the geometric cross-sectional area projected onto a plane perpendicular to the incident beam (i.e., $\sigma_g = \pi a^2$ for a spherical particle). It should be noted that the anisotropy value, g, ranges from 0 to 1, where $g = 0$ represents isotropic scattering and $g = 1$ represents only forward scattering.

As an example, consider a scattering particle to be a polystyrene microsphere suspended in water. The refractive indices of the polystyrene microsphere and the surrounding medium, n_2 and n_1, are 1.59 and 1.33, respectively. The scattering efficiency, Q_s, and the anisotropy value, g, as a function of the relative particle size, A, are shown in Fig. 2.3 [11]. The scattering efficiency, Q_s, increases rapidly for a

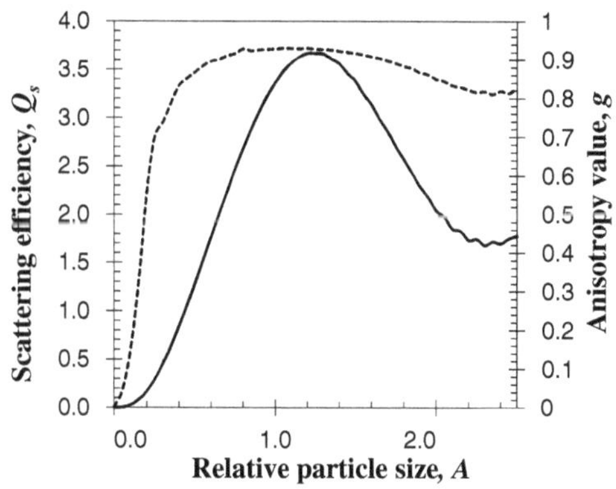

Fig. 2.3 Scattering efficiency, Q_s (*solid curve*) and anisotropy value, g (*dashed curve*) as a function of the relative particle size, A ($n_1 = 1.33$ and $n_2 = 1.59$). Reprinted with permission from [11], Cambridge University Press

relative particle size, A, between 0.2 to 1.2, and then reaches the maximum value when the relative particle size, A, is approximately 1.25. The anisotropy value, g, increases rapidly when the relative particle size, A, is less than 0.4, and reaches a fairly steady region when the relative particle size, A, is larger than 0.5.

2.2.3 Scattering Mean Free Path Length

Another important scattering parameter is the scattering mean free path (smfp) length, l_s, when a large number of scatterers are considered. When a beam of light propagates through such a scattering medium, photons experience scattering. The distance within which a photon experiences no scattering is called the scattering free path. Because scatterers are randomly dispersed in a turbid medium, the scattering-free-path length varies in each scattering event. The scattering mean free path length, l_s is a measure of the average free-path-length that a photon travels before it is scattered. The scattering mean free path length, l_s, is determined via

$$l_s = \frac{1}{\sigma_s \rho}, \tag{2.11}$$

where ρ is the concentration of particles per cubic micrometer and is given by

$$\rho = \frac{6W}{p\pi d^3}. \tag{2.12}$$

Here p is the density of the scattering particles, W is the weight percentage of scatterers in a bulk solution and d is the diameter of scatterers.

Table 2.1 The calculated scattering mean free path length, l_s, and the anisotropy value g, for the polystyrene microspheres of different diameters for an excitation wavelength, λ, of 632.8 nm

Sphere diameter, κ (μm)	Geometric cross-section, σ_g (μm^2)	Concent. ρ (part./μm^3)	Relative particle size, A (a/λ)	Scattering efficiency, $Q_s = \sigma_s/\sigma_g$	Scattering cross-section, σ_s (μm^2)	smfp length, l_s (μm)	Anisotropy value, g
0.107	0.009	37.11	0.0845	0.0037	7.19×10^{-3}	416.4	0.076
0.202	0.032	5.52	0.1596	0.0805	2.58×10^{-3}	70.2	0.31
0.48	0.181	0.411	0.3793	0.7002	1.27×10^{-1}	19.2	0.81
1.056	0.876	0.038	0.8342	2.7784	2.42	10.8	0.92

For a turbid medium consisting of polystyrene microspheres suspended in water, its standard weight percentage W is 2.5 %. Table 2.1 shows the parameters for the four types of polystyrene microspheres used in Chaps. 4 and 5 for a wavelength, λ, of 632.8 nm. From the results illustrated in Table 2.1 it is seen that larger polystyrene microspheres have shorter scattering-mean free path lengths and higher anisotropy values. This result indicates that increasing the scattering microsphere diameter, d, for a given sample geometric thickness and a given weight percentage of a scattering solution, has the effect of increasing the average number of scattering events experienced by an individual photon since the free-path-length is significantly shorter. An increase in the scattering microsphere diameter, d, also significantly changes the angular distribution of the scattered photons, since more scattered photons travel in the forward direction as the anisotropy value g, approaches unity. These two effects play an important and significant role in determining the success of a gating mechanism.

2.3 Stokes Vector

Based on the Mie scattering theory, one can derive the polarization state of the incident light before and after each scattering event. Consider a monochromatic plane wave incident upon a spherically symmetric scatterer. The polarization state of the incident light can be represented by the Stokes vector [3]

$$I_{\text{incident}} = \begin{bmatrix} I_{\text{inc}} \\ Q_{\text{inc}} \\ U_{\text{inc}} \\ V_{\text{inc}} \end{bmatrix}, \tag{2.13}$$

where

$$I_{\text{inc}} = \left\langle E_x E_x^* + E_y E_y^* \right\rangle, \tag{2.14}$$

$$Q_{\text{inc}} = \left\langle E_x E_x^* - E_y E_y^* \right\rangle, \tag{2.15}$$

$$U_{\text{inc}} = \left\langle E_x E_y^* + E_y E_x^* \right\rangle, \tag{2.16}$$

$$V_{\text{inc}} = \left\langle E_x E_y^* - E_y E_x^* \right\rangle. \tag{2.17}$$

Here E_x and E_y represent the orthogonal components of an incident electric field along the x and y axes. $\langle\rangle$ denotes the time average of the bracketed quantity.

Note that in an optical microscope, the illumination beam on a sample is a convergent beam from an objective lens. Assume that the incident beam before an illumination objective is a linearly polarized plane wave. The relationship of the polarization state between the incident light and the scattered light can be represented as [12, 13]

$$I_{\text{scat}}^{\text{after}} = M I_{\text{scat}}^{\text{before}}, \tag{2.18}$$

where $I_{\text{scat}}^{\text{before}}$ and $I_{\text{scat}}^{\text{after}}$ are the Stokes vectors before and after the scattering event, respectively, and M is a 4×4 scattering matrix for a spherical symmetric particle [3]:

$$M = \begin{bmatrix} S_{11}(\theta) & S_{12}(\theta) & 0 & 0 \\ S_{12}(\theta) & S_{11}(\theta) & 0 & 0 \\ 0 & 0 & S_{33}(\theta) & S_{34}(\theta) \\ 0 & 0 & -S_{34}(\theta) & S_{33}(\theta) \end{bmatrix}, \tag{2.19}$$

where the matrix elements are given by

$$S_{11}(\theta) = \frac{1}{2}\left(|S_2(\theta)|^2 + |S_1|(\theta)^2\right), \tag{2.20}$$

$$S_{12}(\theta) = \frac{1}{2}\left(|S_2(\theta)|^2 - |S_1(\theta)|^2\right), \tag{2.21}$$

$$S_{33}(\theta) = \frac{1}{2}\left(S_2^*(\theta)S_1(\theta) + S_2(\theta)S_1^*(\theta)\right), \tag{2.22}$$

$$S_{34}(\theta) = \frac{1}{2}\left(S_1(\theta)S_2^*(\theta) - S_2(\theta)S_1^*(\theta)\right). \tag{2.23}$$

Here

$$S_1(\theta) = \sum_{i=1}^{\infty} \frac{2i+1}{i(i+1)}(a_i\pi_i + b_i\tau_i), \tag{2.24}$$

$$S_2(\theta) = \sum_{i=1}^{\infty} \frac{2i+1}{i(i+1)} (a_i \tau_i + b_i \pi_i), \tag{2.25}$$

where a_i and b_i are the scattering coefficients defined by (2.3) and (2.4), respectively, and π_i and τ_i are angle-dependent functions defined as [3]

$$\pi_i = \frac{P_i^1}{\sin\theta}, \tag{2.26}$$

$$\tau_i = \frac{\mathrm{d}P_i^1}{\mathrm{d}\theta}, \tag{2.27}$$

where P_i^1 is the associated Legendre function [10]. Both π_i and τ_i can be calculated from the upward recursion relations given by [3]

$$\pi_i = \left(\frac{2i-1}{i-1}\pi_{i-1}\right)\cos\theta - \left(\frac{i}{i-1}\pi_{i-2}\right), \tag{2.28}$$

and

$$\tau_i = (i\cos\theta)\pi_i - (i+1)\pi_{i-1}, \tag{2.29}$$

where $\pi_0 = 0$ and $\pi_1 = 1$.

References

1. W.F. Cheong, S.A. Prahl, A.J. Walsh, A review of the optical properties of biological tissues. IEEE J. Quantum Electron. **26**, 2166 (1990)
2. S.P. Morgan, M.P. Khong, M.G. Somekh, Effects of polarization state and scatterer concentration on optical imaging through scattering media. Appl. Opt. **36**, 1560 (1997)
3. C.F. Bohern, D.R. Huffman, *Absorption and Scattering of Light by Small Particles* (Wiley, New York, 1983)
4. A. Ishimaru, *Wave Propagation and Scattering in Random Media* (Academic, New York, 1978)
5. M. Born, E. Wolf, *Principles of Optics* (Pergamon, New York, 1980)
6. A. Wax, C.H. Yang, V. Backman, K. Badizadegan, C.W. Boone, R.R. Dasari, M.S. Feld, Cellular organization and substructure measured using angle-resolved low-coherence interferometry. Biophys. J. **82**, 2256 (2002)
7. A. Wax, C. Yang, M. Müller, R. Nines, C.W. Boone, V.E. Steele, G.D. Stoner, R.R. Dasari, M.S. Feld, In situ detection of neoplastic transformation and chemopreventive effects in rat esophagus epithelium using angle-resolved low-coherence interferometry. Cancer Res. **63**, 3556 (2003)
8. R. Jullien, R. Botet, *Aggregation and Fractal Aggregates* (World Scientific, Singapore, 1987)
9. X. Deng, X. Gan, M. Gu, Effective Mie scattering of a spherical aggregate and its application in turbid media. Appl. Opt. **43**, 2925 (2004)

10. M. Abramowitz, I.A. Stegun, *Handbook of Mathematical Functions* (Dover Publications Inc., New York, 1970)
11. X. Gan, S. Schilders, M. Gu, Combination of annular aperture and polarisation gating methods for efficient microscopic imaging through a turbid medium: theoretical analysis. Microsc. Microanal. **3**, 495 (1997)
12. X. Gan, S. Schilders, M. Gu, Image enhancement through turbid media under a microscope using polarization gating methods. J. Opt. Soc. Am. A **16**, 2177 (1999)
13. J.M. Schmitt, A.H. Gandjbakhche, R.F. Bonner, Use of polarised light to discriminate short-path photons in a multiply scattering medium. Appl. Opt. **31**, 6535 (1992)

Chapter 3
Monte Carlo Simulation for an Optical Microscope

Abstract This chapter covers the basic principles that are used in Monte Carlo simulation for investigation into microscopy imaging through tissue media. Section 3.1 summarizes the basic formula in a conventional Monte Carlo simulation process. The implementation of this method in reflection and transmission optical microscopes is given in Sect. 3.2. The effect of the polarization states and pulsed illumination of a beam are described in Sects. 3.3 and 3.4. Sections 3.5–3.8 are dedicated to dealing with the various features of turbid media including the effect of the boundary, scatterer size, and aggregation. In Sects. 3.9 and 3.10, Monte Carlo simulation methods for multi-photon fluorescence and coherent imaging processes are discussed.

3.1 Model of Monte Carlo Simulation

Monte Carlo simulation is an important tool in studying photon propagation through turbid media since it allows the efficient tracking of photon states (direction, polarization, etc.) at each scattering event experienced within turbid media [1–3].

In Sect. 2.2, the scattering mean free path length, l_s, has been defined as a measure of the average free path length before a photon is scattered. The decay in the number of unscattered photons after propagating through a depth, d, in turbid media, for a collimated beam, can be represented by Beer's law [4] and is equal to the probability $P(d)$ that a photon has a free path length, d, for a given scattering mean free path length, l_s,

$$P(d) = N(d)/N(0) = \exp(-d/l_s), \tag{3.1}$$

where $N(d)$ is the number of the unscattered photons at depth d, and $N(0)$ is the number of incident photons [3].

For a given number of photons incident to a turbid medium the free path length between two consecutive interactions of the photons is given as

M. Gu et al., *Microscopic Imaging Through Turbid Media*,
Biological and Medical Physics, Biomedical Engineering,
DOI 10.1007/978-3-662-46397-0_3

$$l_t = -l_s \ln \beta \tag{3.2}$$

where $l_s = \langle l_t \rangle$ is the time averaged scattering mean free path length, and β is a random number which is evenly chosen between 0 and 1 and satisfies the probability function in (3.1).

In the Monte Carlo model developed for microscopic imaging systems [3], five independent variables are needed in order to simulate a multi-dimensional photon distribution; these include three spatial coordinates x, y, and z and two angular coordinates (a polar angle, θ and an azimuthal angle, ϕ). Figure 3.1 illustrates the coordinate system for a spherical scattering particle. These five independent variables are not necessarily set to zero for a microscopic imaging system and the method used for determining their initial values is outlined in Sect. 3.2.

Each time a photon interacts with a scattering particle in the simulation model, new coordinates are determined. The new polar angle, θ, is equated to the scattering angle, θ_s, which is chosen based on the Henyey–Greenstein (H–G) probability distribution [5]

$$P_\theta(\theta) = \frac{1 - g^2}{2(1 + g^2 - 2g\cos\theta)^{3/2}} \sin\theta, \tag{3.3}$$

where g is the anisotropy value, calculated from (2.9). The scattering polar angle, θ_s, can therefore be obtained from [5]

$$\cos\theta_s = \frac{1 + g^2}{2g} - \frac{(1 - g^2)^2}{2g(1 - g + 2g\delta)^2}, \tag{3.4}$$

where δ is another evenly distributed random number from 0 to 1. The new azimuthal angle, ϕ, is randomly chosen between 0 and 2π, while the spatial coordinates, x, y, and z are determined by the sample and illumination conditions. In Sect. 3.3, we will describe the detail of photons migration through a layer of a turbid medium.

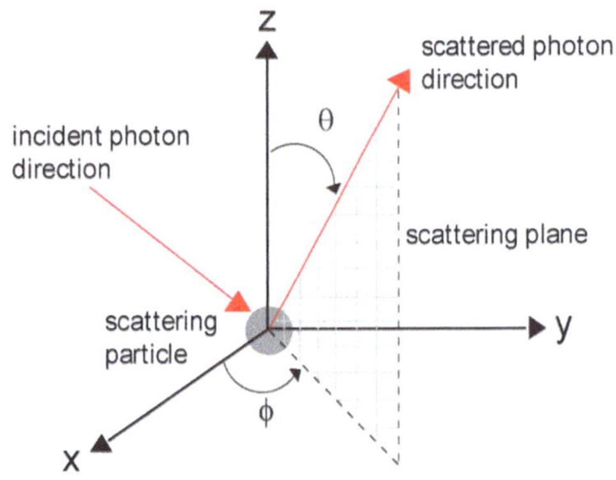

Fig. 3.1 Light scattering by a spherical particle

3.2 Microscopic Imaging

Because of the angular distribution of propagating photons through an objective in a microscope (see Fig. 3.2), the initial values of the three spatial coordinates x, y, and z and two angular coordinates θ and ϕ for a photon entering a turbid medium are not necessarily set to be 0. Figure 3.2 illustrates the geometric arrangement

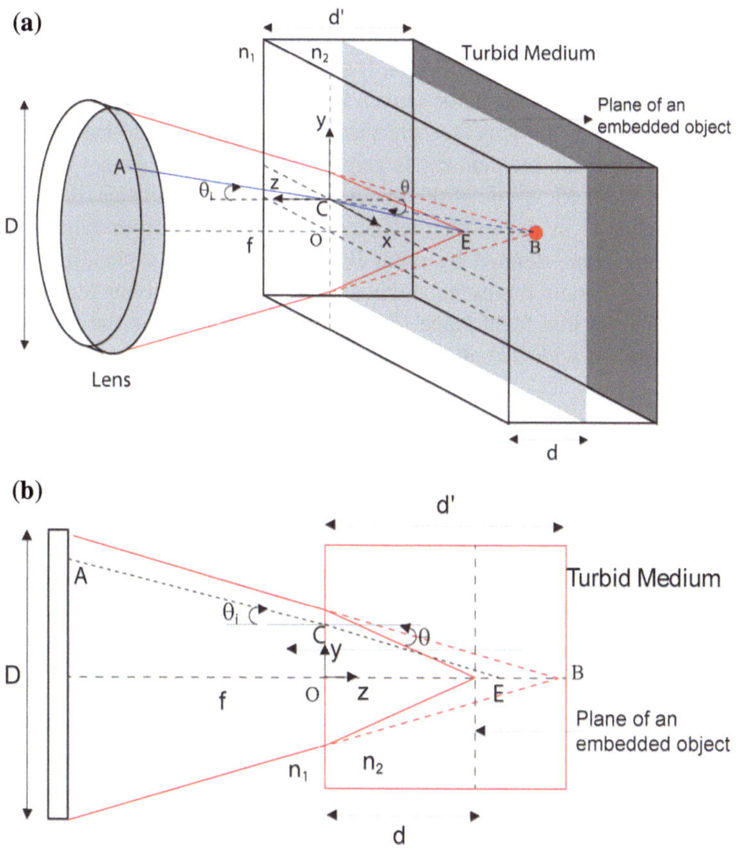

Fig. 3.2 Schematic for photon migration through an objective. **a** Three-dimensional view; **b** two-dimensional view in the meridian plane. The full projection space for photons is illustrated in *black*, while the incident projection of a single photon is given in *gray*. The figure illustrates the case when $n_1 > n_2$

between the incident photons originating from a thin lens and a turbid medium. The origin, O, of the x–y–z coordinate system is defined as the point on the outer surface of the turbid medium that aligns through the focus and the center of the lens. With this convention the penetration depth, d', into the turbid sample can be directly determined.

The x–y coordinate of the starting point A of the incident ray in the lens aperture is determined by a normalized two-dimensional circle random number generator which is scaled to the aperture diameter, D, of the lens [6]. The z-coordinate of the starting point A is given by the difference between the penetration depth, d', and the focal length, f, of the lens, which is derived for a given lens of numerical aperture (NA) and aperture diameter, D (i.e. $f \approx D/2\text{NA}$). It should be noted that the z-coordinate of point A is a constant less than 0, for a given lens, since all photon rays originate from the same x–y plane in the lens (see the shaded gray plane on the lens in Fig. 3.2). The ending point of the incident ray is given by point B, with its x–y–z coordinate defined as the geometric focus of the imaging lens (i.e., coordinates (0, 0, d')).

Due to a change in the refractive index between the surrounding and turbid media, the focal plane defined by point B (see the dark gray plane in the turbid medium in Fig. 3.2) only represents the imaginary depth, d', into the turbid medium. This means that the true focal plane, d, defined by point E, into the turbid sample must be determined with the aid of Snell's law [4]:

$$n_1 \sin \theta_\text{i} = n_2 \sin \theta, \tag{3.5}$$

where n_1 and n_2 are the refractive indices of the surrounding and turbid media, respectively. θ_i is the incident angle at which the ray of a projected photon, defined by points A and B, intersects with the turbid medium boundary at position C (see Fig. 3.2). It should be noted that in the Monte Carlo model spherical aberration is ignored. The convergent focal spot is determined from the maximum incident angle, θ_i, i.e., the marginal ray of the imaging lens.

Reflection and transmission on the interface between a turbid medium and its surrounding medium (see point C in Fig. 3.2) are also considered in this simulation. A weighting factor, f_p, is assigned to each photon to represent its contribution weighting. Initially the weighting factor, f_p, is equal to 1. When a photon passes through a dielectric interface, the weighting of the photon in the transmitted (T) and reflected (R) directions are decided according to the Fresnel formulae [4]

$$T_\text{p} = \frac{2n_1 \cos \theta_\text{i}}{n_2 \cos \theta_\text{i} + n_1 \cos \theta}, \tag{3.6}$$

$$T_\text{s} = \frac{2n_1 \cos \theta_\text{i}}{n_1 \cos \theta_\text{i} + n_2 \cos \theta}, \tag{3.7}$$

$$R_\text{p} = \frac{n_2 \cos \theta_\text{i} - n_1 \cos \theta}{n_2 \cos \theta_\text{i} + n_1 \cos \theta}, \tag{3.8}$$

$$R_s = \frac{n_1 \cos\theta_i - n_2 \cos\theta}{n_2 \cos\theta_i + n_1 \cos\theta}, \tag{3.9}$$

where the subscripts p and s represent parallel and perpendicular polarization components, respectively, while subscripts 1 and 2 represent the surrounding and turbid media, respectively.

The initial values of the five parameters for a single photon (ray) incident to the turbid medium at point C are therefore obtained as follows: the spatial coordinates, x and y, and the angular coordinate, ϕ, are derived from the geometry in Fig. 3.2, while θ is obtained from Snell's law (3.5). The spatial coordinate, z, is set to zero since all photons originate from the same boundary surface (z-plane) of the turbid medium. It should be noted that for transillumination modeling the values of z and θ are initially set to zero, since a parallel beam probe with respect to the turbid media is used.

Two typical scanning optical microscope arrangements used in Chaps. 5 and 6 can be simulated using the Monte Carlo model in Sect. 3.1. They are the transmission scanning optical microscope (Fig. 3.3a) and the reflection scanning optical microscope (Fig. 3.3b). For the transmission optical microscope illustrated in Fig. 3. 3a, a parallel beam is incident upon the imaging lens L_1, which has its focus on the embedded object. A separate collection lens L_2 which has its focus overlapped with

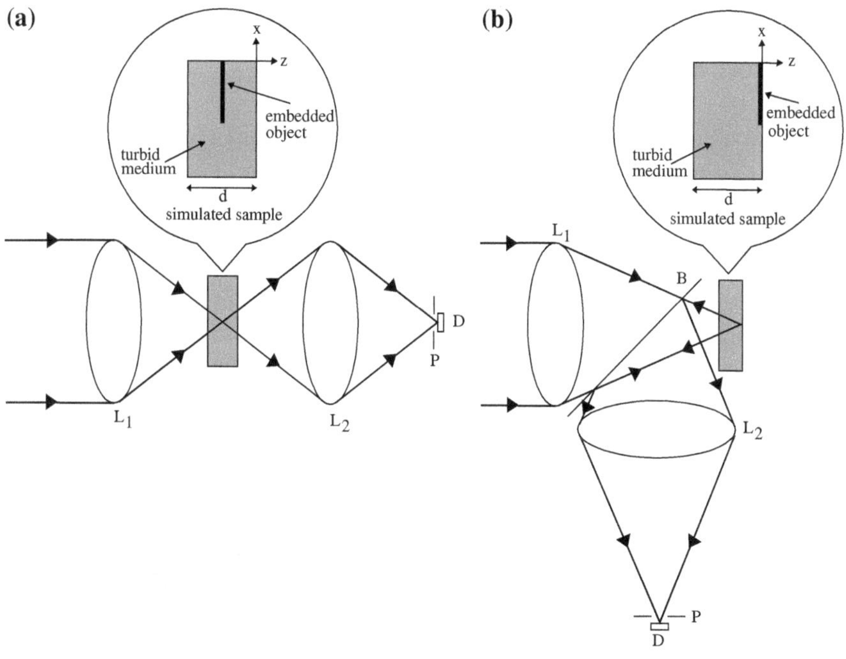

Fig. 3.3 Schematic diagram of the modeled transmission (**a**) and reflection (**b**) scanning optical microscopes. *B* beamsplitter; *D* detector; *L* lenses; *P* pinhole

the imaging lens L_1 is used to collect the light originating from a simulated turbid medium. The collection lens L_2 is also used to focus the collected light onto the detector D. It should be noted that only the photons that fall within the collection cone of lens L_2 will be traced to the detector plane by geometric optics. The simulated embedded object is placed at the center of the turbid medium in this optical arrangement. The simulated samples embedded within the turbid medium for the transmission arrangement are considered to have 100 % absorption. That is, when an incident photon hits the embedded object it ceases to be simulated. For the reflection microscope illustrated in Fig. 3.3b a parallel beam is incident upon the imaging lens L_1 which has its focus on an embedded object. A collection lens L_2 then focuses the collected light originating from the turbid medium onto the detector D. It should be noted that only the photons that fall within the collection cone of lens L_2 are traced to the detector plane by geometric optics. In this arrangement the embedded object is placed on the back surface of the turbid medium. The simulated samples embedded within the turbid medium for the reflection arrangement are considered to have 100 % reflectivity. With this arrangement the reflection microscope is equivalent to folding the transmission microscope arrangement, at the embedded object, onto itself. It should be noted that the size of the detector used in the detection of the light originating from the simulated turbid media is determined by the effective size of the used pinhole, P (see Fig. 3.3).

3.3 Effect of Polarization

According to the Stokes vector theory in Sect. 2.3, once the scattering parameters, $S_1(\theta)$ and $S_2(\theta)$ in (2.24) and (2.25) are known, all elements of the scattering matrix, M, defined in (2.18) and (2.19), can be calculated. Thus the polarization gating mechanism that will be discussed in Chap. 6 can be modeled using the Monte Carlo simulation method.

Since calculating the scattering parameters, $S_1(\theta)$ and $S_2(\theta)$, is a time-consuming task, an approximation is used in the calculation of the scattering matrix, M, at each scattering event. Before simulation starts [7], the scattering angle, θ, between 0 and π is first divided into 1,000 equal intervals and the scattering parameters, $S_1(\theta)$ and $S_2(\theta)$, corresponding to those scattering angles are calculated. All the calculated scattering angles, θ, and the scattering parameters, $S_1(\theta)$ and $S_2(\theta)$, are stored in a database file. To employ polarization gating in the Monte Carlo simulation, Stokes parameters are defined for each incident photon in addition to the five existing parameters (x, y, z, θ and ϕ). At each scattering event in the Monte Carlo simulation [7] the scattering parameters, $S_1(\theta)$ and $S_2(\theta)$, are obtained by matching the scattering angle, θ_s (determined by the Henyey–Greenstein probability distribution defined in (3.4)), with the closest scattering angle, θ, stored in the lookup database file. Then according to the chosen scattering angle, θ, and its corresponding scattering parameters, $S_1(\theta)$ and $S_2(\theta)$, all the elements of the scattering matrix, M, are calculated using (2.19)–(2.25). Therefore, a new Stokes vector can be calculated from (2.18).

By monitoring the change of the polarization state of each photon at every scattering event, the depolarization of the light propagating through a turbid medium can be evaluated. The degree of polarization, γ, of the detected scattered light in this book is defined as

$$\gamma = \frac{I_p - I_s}{I_p + I_s}, \tag{3.10}$$

where I_p and I_s are the light intensity detected with the analyzer parallel and perpendicular to the incident polarization direction, respectively. Assuming that the incident light is linearly polarized in the x-direction, the parallel polarization intensity, I_p, and the perpendicular polarization intensity, I_s, can be calculated as

$$I_p = I_{scat} \cdot \begin{bmatrix} 1 & 1 & 0 & 0 \end{bmatrix}, \tag{3.11}$$

and

$$I_s = I_{scat} \cdot \begin{bmatrix} 1 & -1 & 0 & 0 \end{bmatrix}. \tag{3.12}$$

The polarization gating mechanisms used in Chap. 6 [7] are parallel (conventional) polarization-gating, based on the signal intensity (I_p) detected with an analyzer parallel to the direction of the incident polarization, and perpendicular polarization-gating, based on the signal intensity (I_s) detected with an analyzer perpendicular to the direction of the incident polarization. The third method of polarization-gating considered is differential polarization gating, which is based on the subtraction of the conventional polarization-gated intensity, I_p, and the perpendicular polarization-gated intensity, I_s [7].

3.4 Effect of Pulsed Illumination

To describe the pulse propagation through a turbid in the Monte Carlo model [3, 8], we need to include the difference of the photon propagation caused by time-of-flight in a pulsed beam. An ultrashort pulsed beam with pulse width, $\Delta\tau_0$, has a finite distribution of wavelength (frequency) components. According to Mie scattering theory (see Sect. 2.2), the scattering coefficients, a_i and b_i, are related to the ratio between the radius of a scattering particle, a, and the wavelength of light, λ (see (2.1) to (2.4)). Therefore, for a given scattering particle radius, a, the scattering coefficients, a_i and b_i, for each wavelength component, λ, are different. Hence the scattering cross-section, σ_s, and the anisotropy value, g, need to be determined for each individual wavelength component, λ, for an illumination pulse.

The effect of the broad spectrum of an ultrashort pulse on the scattering efficiency, Q_s, and the anisotropy value, g, becomes more pronounced as the pulse width, $\Delta\tau_0$, becomes shorter. Let us consider that the intensity of an illumination source is a Gaussian-shaped pulse given as [9]

$$U_0(t) = \exp(-i\omega_0 t)\exp\left[-\left(\frac{t}{T}\right)^2\right], \tag{3.13}$$

where t is the local time coordinate, T is related to the pulse width $\Delta\tau_0$ via $\Delta\tau_0 = 2T\sqrt{\ln 2}$, and ω_0 is the central frequency. The corresponding Fourier spectrum for the Gaussian-shaped pulse is then given as [9]

$$V_0(\Delta\Omega) = \sqrt{\pi}T\exp\left[-\left(\frac{T\Delta\Omega}{2}\right)^2\right], \tag{3.14}$$

where $\Delta\Omega$ is the spectral width which is defined as the total bandwidth between two positions at which the intensity drops to one half of its peak value. The relationship between the pulse width $\Delta\tau_0$, and the spectral width $\Delta\Omega$ for the pulse is then given as [8]

$$\Delta\Omega = \frac{8\ln 2}{\Delta\tau_0}. \tag{3.15}$$

For example, a 10-fs pulse with a central wavelength, λ_0 of 700 nm, has a corresponding spectral width, $\Delta\Omega$, of approximately 0.206. Note that in this case the spectral width, $\Delta\Omega$, is normalized by the central frequency, ω_0 ($\omega_0 = 2\pi c/\lambda_0$, where $c = 3 \times 10^8$ m/s).

To understand the effect of the spectral width, $\Delta\Omega$, on the scattering efficiency, Q_s, and the anisotropy value, g, we define two parameters,

$$\Delta Q_s = Q_s(\omega + 0.5\Delta\Omega) - Q_s(\omega - 0.5\Delta\Omega), \tag{3.16}$$

and

$$\Delta g = g(\omega + 0.5\Delta\Omega) - g(\omega - 0.5\Delta\Omega). \tag{3.17}$$

Since the pulse width, $\Delta\tau_0$, can be directly related to the frequency bandwidth, $\Delta\Omega$ we can plot the difference in the scattering efficiency, ΔQ_s and the difference in the anisotropy value, Δg, as a function of the pulse width $\Delta\tau_0$. Figure 3.4 is the effect of the pulse width, $\Delta\tau_0$, on the difference in the scattering efficiency, ΔQ_s and the difference in anisotropy value, Δg, for a central wavelength, λ_0, of 700 nm and a scattering particle radius, a, of 0.518 μm [3]. It is seen that the difference in the scattering efficiency, ΔQ_s, and the difference in the anisotropy value, Δg, both increase as the pulse width, Δt_0, decreases. For the pulse width, $\Delta\tau_0$, less than 20 fs the difference in the scattering efficiency, ΔQ_s, and the difference in the anisotropy

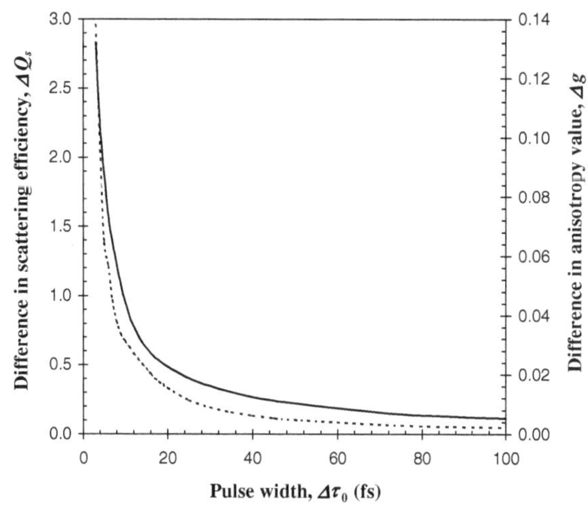

Fig. 3.4 Difference in the scattering efficiency, ΔQ_s, (*solid curve*) and the difference in the anisotropy value, Δg, (*dashed curve*) as a function of the pulse width, $\Delta\tau_0$ ($n_1 = 1.33$ and $n_2 = 1.59$)

value, Δg, become more pronounced, indicating that the effect of the frequency bandwidth is more significant in this region.

With the help of Fig. 3.4, the illumination of an ultrashort pulse can be incorporated into the Monte Carlo simulation by describing that the temporal distribution of incident photons is given by the following Gaussian profile [3]:

$$f(t_0) = \exp\left(-\frac{t_0^2}{T^2}\right), \tag{3.18}$$

where t_0 is the departure time of a photon. The departure time, t_0, of photons in the pulse is randomly chosen according to the temporal distribution in (3.18).

3.5 Photon Migration Through a Layer of a Turbid Medium

In the case of Monte Carlo simulation of photon propagation through a layer of a scattering medium, photons are simply treated as particles without any wave feature. Individual photons suffer events of scattering and absorption, which is related to the local optical properties of the turbid medium characterized by scattering coefficient μ_s and absorption coefficient μ_a (Fig. 3.5). This $\mu_t = \mu_s + \mu_a = 1/l_s$. The Monte Carlo simulation is performed by a numerical step-by-step tracing of the random migration of the photons within the scattering medium until their annihilation or escapement [10, 11].

It is clear that each step a photon may take from the position $(\vec{P}_{n-1})$ to the next position $(\vec{P}_n)$ can be easily traced by three random variables in a three-dimensional

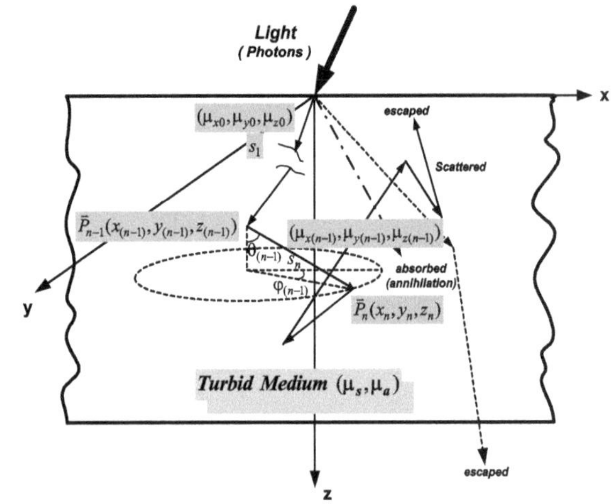

Fig. 3.5 Schematic diagram of the Monte Carlo simulation in dealing with the photon propagation through a later of *s* scattering medium

dynamic spherical coordinate system according to Fig. 3.5, they are the step length (s), the deflection angle ($0 \leq \theta < \pi$), and the azimuthal angle ($0 < \varphi < 2\pi$), which, respectively, describe the distance a photon takes between two successive photon–medium interaction sites, the angle of deflection between the photon propagation direction from the z axis and the corresponding azimuthal angle when a scattering event occurs.

The value distributions of the variables such as the step length (s) and the scattering angles (θ, φ) are determined by their corresponding physically described probability density functions $p(s)$, $p(\theta)$, $p(\varphi)$, based on the following sampling rules:

$$\begin{cases} \int_a^b p(x) = 1 & \text{for} \quad x \in [a, b] \\ \int_a^{x_1} p(x) = \xi_1 & \text{for} \quad \xi_1 \in [0, 1] \end{cases} \tag{3.19}$$

where x represents any of the random variables s, θ, or φ and ξ_1 are one of the uniformly distributed random numbers within 0–1.

The probability function $p(s)$ can be derived Beer's law and the Henyey and Greenstein function, as shown in Sect. 3.1. Therefore, the choice for s, $\cos\theta$ and φ can be expressed as

$$s = -l_s \ln(\xi) = -\, l_s \ln(1 - \xi) \tag{3.20}$$

$$\cos\theta = \begin{cases} \frac{1}{2g}\left\{1 + g^2 - \left[\frac{1-g^2}{1-g+2g\zeta}\right]^2\right\}, & g \neq 0, \\ 2\zeta - 1, & g = 0, \end{cases} \tag{3.21}$$

and

$$\varphi = 2\pi\xi, \tag{3.22}$$

respectively. Here ξ is the uniformly distributed random number within 0–1.

The moving of the photon packets in the global Cartesian coordinate system can be described as follows. Suppose that a photon packet locates at a current position $\vec{P}_{n-1}(x_{(n-1)}, y_{(n-1)}, z_{(n-1)})(n \geq 1)$ in the global coordinate system, the new position of the photon packet along its trajectory direction in the global coordinate $\vec{P}_n(x_n, y_n, z_n)$ can be updated as [10, 17]

$$\begin{cases} x_n = x_{(n-1)} + \mu_{x(n-1)} s_n, \\ y_n = y_{(n-1)} + \mu_{y(n-1)} s_n, \\ z_n = z_{n-1} + \mu_{z(n-1)} s_n, \end{cases} \tag{3.23}$$

where the incident directional cosines $(\mu_{x0}, \mu_{y0}, \mu_{z0})$ can be calculated as

$$\begin{cases} \mu_{x0} = \sin\theta_0 \cos\varphi_0, \\ \mu_{y0} = \sin\theta_0 \sin\varphi_0, \\ \mu_{z0} = \cos\theta_0, \end{cases} \tag{3.24}$$

where θ_0 and φ_0 are the deflection and azimuthal angles of the incident beam entering the turbid medium and s_1 is the first scattering step size along the direction $(\mu_{x0}, \mu_{y0}, \mu_{z0})$.

After a new scattering event, once the scattering angles θ_n and φ_n have been determined according to the sampling mentioned in the above section, the new directional cosines $(\mu_{xn}, \mu_{yn}, \mu_{zn})$ of this photon packet in the global coordinate system can be calculated as [10, 11]

$$\begin{cases} \mu_{xn} = \frac{\sin\theta_n}{\sqrt{1-\mu^2_{z(n-1)}}}\left(\mu_{x(n-1)}\mu_{z(n-1)}\cos\varphi_n - \mu_{y(n-1)}\sin\varphi_n\right) + \mu_{x(n-1)}\cos\theta_n, \\ \mu_{yn} = \frac{\sin\theta_n}{\sqrt{1-\mu^2_{z(n-1)}}}\left(\mu_{y(n-1)}\mu_{z(n-1)}\cos\varphi_n - \mu_{x(n-1)}\sin\varphi_n\right) + \mu_{y(n-1)}\cos\theta_n, \\ \mu_{zn} = -\sin\theta_n\cos\varphi_n\sqrt{1-\mu^2_{z(n-1)}} + \mu_{z(n-1)}\cos\theta_n. \end{cases} \tag{3.25}$$

If $\left|\mu_{z(n-1)}\right| > 0.99999$, then the following formulas should be used:

$$\begin{cases} \mu_{xn} = \sin\theta_n \cos\varphi_n, \\ \mu_{yn} = \sin\theta_n \sin\varphi_n, \\ \mu_{zn} = \cos\theta_n \ (\text{if } \mu_{z(n-1)} > 0) \text{ or } \mu_{zn} = -\cos\theta_n \ (\text{if } \mu_{z(n-1)} < 0). \end{cases} \tag{3.26}$$

3.6 Effect of Multiple Layers

When a photon passes through a dielectric interface between two layers of turbid media, this photon is either reflected or transmitted. Snell's law indicates the relationship between the angle of incidence, θ_i, the angle of transmittance, θ_t, and the refractive indices at the media that the photon is incident from, n_i, and transmitted to n_t. According to the Fresnel's formulas in (3.6)–(3.9), the reflection and the transmission coefficients of a beam with parallel and perpendicular polarizations can be determined. The reflectance at the angle of incidence $\theta_i, R(\theta_i)$, can be calculated as

$$R(\theta_i) = \frac{1}{2}\left(|R_\perp|^2 + |R_{//}|^2\right) = \frac{1}{2}\left[\frac{\sin^2(\theta_i - \theta_t)}{\sin^2(\theta_i + \theta_t)} + \frac{\tan^2(\theta_i - \theta_t)}{\tan^2(\theta_i + \theta_t)}\right]. \quad (3.27)$$

In the Monte Carlo simulation, a photon is either reflected or transmitted on the boundary. The probability of reflection or transmission of a photon by the boundary is determined by a uniformly distributed random number between 0 and 1, ξ, which is compared to $R(\theta_i)$. If $\xi \leq R(\theta_i)$, the photon is internally reflected (Fig. 3.6). If $\xi > R(\theta_i)$, the photon escapes from the current layer of tissue (Fig. 3.10).

To determine if a photon hits the boundary during each step s of the propagation of a photon, one can use the following method. After the step size s and the scattering angles θ and φ are decided, the next position where the photon may arrive from the current position (x, y, z) as if there were no boundary should be calculated. Then, the following condition applies:

$$\begin{cases} \text{if} \quad \mu_z > 0 \quad \text{and} \quad z \leq z_1 & \text{hits the lower boundary of the current layer,} \\ \text{if} \quad \mu_z = 0 & \text{discards the photon,} \\ \text{if} \quad \mu_z < 0 \quad \text{and} \quad z \leq z_0 & \text{hits the upper boundary of the current layer,} \end{cases}$$

where z_0 and z_1 are the z coordinates of the upper and lower boundaries of the current layer.

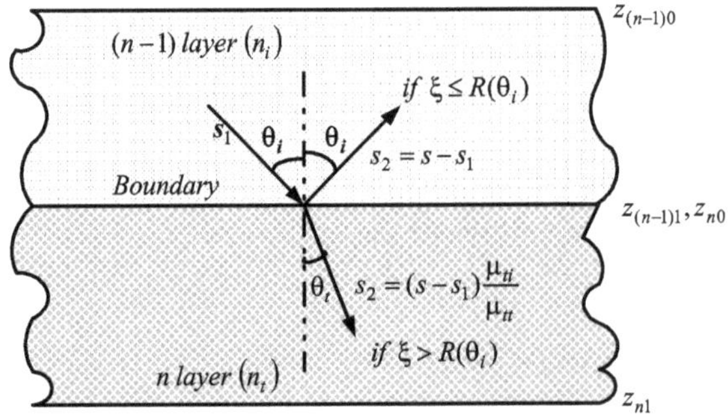

Fig. 3.6 A photon is reflected or transmitted at the interface between two layers of turbid media

The next step is to determine the distance between the current position and the boundary. Suppose that the step size of the photon can freely take in the current medium is s as if there were no boundary. The distance of the photon propagates before it meets a boundary is s_1 along the direction of the photon propagation, which can be calculated as $|(z - z_0)/\mu_z|$ or $|(z - z_1)/\mu_z|$ depending on whether the upper or the lower boundary is met. Once it meets the boundary and ($s_1 < s$), if it is reflected back into the same layer of medium, the photon can continue its propagation along the reflection direction with an angle of reflection, the same as the angle incidence θ_i in the first medium and continuously propagate with the remaining step size s_2, which can be simply determined as $s_2 = s - s_1$. If the photon escapes from the first medium (μ_{ti}) and transmits to the second medium (μ_{tt}), the photon propagates along the transmission direction with the angle of transmission θ_t and continues to propagate with the remaining step size s_2, which is related to the relative optical properties of the two media and can be described as

$$s_2 = (s - s_1)\frac{\mu_{ti}}{\mu_{tt}} \tag{3.28}$$

If $s_1 = s$, a new scattering event should be generated on the boundary.

3.7 Effect of Multi-sized Scatterers

So far, the model for turbid media considered in the Monte Carlo simulation is based on a homogeneous turbid media structure where the size of scatterers is equal. Biological tissue usually exhibits a complex inhomogeneous structure; the basic constitution components of biological tissue, cells, have a wide range of size from nanometers to tens of micrometers and there are numerous different sizes of scattering organelles existing within the cells [12]. For example, most animal cells range from 10 to 30 μm in size and the nuclei, the largest scattering organelles existing in cells, have size ranging between 3 and 10 μm. Mitochondria are small organelles with size ranging from 0.5 to 1.5 μm. Other smaller cell components include endoplasmic reticulum (0.2–1 μm), lysomes (0.2–0.5 μm), and peroxisomes (0.2–0.5 μm) [13]. Therefore, it is necessary to include the effect of multi-sized scatters on the Monte Carlo simulation model.

According to the Mie scattering theory, the size of the scatterers together with other optical properties such as their refract index determines the scattering properties; the cross section and the anisotropy value g. For the sake of clarity and simplicity, let us consider a medium consisting only of two types of particles, which, respectively, have the scattering coefficients μ_1 and μ_2, corresponding to the mean free path lengths l_{s1} and l_{s2}. The probability density distribution $p(l_{s1}, l_{s2})$ for the shorter mean free path length l_{sm} in this medium can be expressed as

$$p(l_{\rm sm}) = \mu_1 e^{-\mu_1 l_{\rm s1}} {\rm d}l_{\rm s1} \cdot \underset{(\text{when } l_{\rm s1} < l_{\rm s2} < \infty)}{\int_{l_{\rm s1}}^{\infty}} \mu_2 e^{-\mu_2 l_{\rm s2}} {\rm d}l_{\rm s2} + \mu_2 e^{-\mu_2 l_{\rm s2}} {\rm d}l_{\rm s2} \cdot \underset{(\text{when } l_{\rm s2} < l_{\rm s1} < \infty)}{\int_{l_{\rm s1}}^{\infty}} \mu_1 e^{-\mu_1 l_{\rm s1}} {\rm d}l_{\rm s1} \quad (3.29)$$

and

$$\int_0^s p(l_{\rm sm}) = 1 - e^{-(\mu_1 + \mu_2)s}, \quad (3.30)$$

which is consistent with the probability density function $p(l_{\rm s})$ in the medium of the mixture of two types of particles, where the total interaction coefficient is $\mu = \mu_1 + \mu_2$. According to the sampling rules, the distribution of the step length s is

$$s = -\frac{\ln(1-\xi)}{\mu_1 + \mu_2} = -l_{\rm s} \ln(1-\xi) \quad \text{and} \quad \frac{1}{l_s} = \frac{1}{l_{\rm s1}} + \frac{1}{l_{\rm s2}}. \quad (3.31)$$

Based on the above probability theory, one can implement the Monte Carlo model for an inhomogeneous turbid medium consisting of n types of spherical scatterers as follows. Suppose that each type of scatterer has a given size (diameter) ρ_i and a concentration c_i in the medium. A scattering mean free path length (l_i) for each type of scatterer is determined by the corresponding concentration (c_i) and scattering cross-section ($\sigma_{\rm si}$), according to the Mie theory [14]:

$$l_{\rm si} = 1/(c_i \sigma_{\rm si}) \quad (i = 1, 2, \ldots, n), \quad (3.32)$$

To determine the real scattering step length between each two consecutive scattering events of a photon in a turbid medium of multi-size scattering particles, we first independently calculate the scattering step length (s_i) from each type of particle according to the equation used in a medium with identical particles [15]:

$$s_i = -l_{\rm si} \ln(\xi) \quad (3.33)$$

where $0 < \xi < 1$ is a randomly produced uniform distribution number. Then the shortest step length (s_m) is taken to be the real length within which a photon can propagate freely. The anisotropy value (g_m) corresponding to the m-th type of particles, which is also calculated based on Mie theory [14], is used to determine the scattering direction of the photon. Meanwhile, for a more straightforward understanding of the image performance of a turbid medium of multi-size scattering particles, the parameter, the effective mean free path length ($l_{\rm s}$), is also introduced as a measure of the randomness in such an inhomogeneous medium:

$$1/l_s = \sum_{i=1}^{n} 1/l_{si} \tag{3.34}$$

The effective mean free path length (l') weights the contributions of scattering cross-sections from different types of particles to the scattering features of the turbid medium. It is an analogous parameter to the mean free path length (l) in a homogeneous medium which has scattering particles of one size.

3.8 Effect of Particle Aggregation

Mie scattering means the interaction of a plane electromagnetic wave with a spherical particle of size comparable to the illumination wavelength [14, 16]. This process can be described by the so-called Mie scattering theory that is based on the diffraction of a plane wave by a spherical particle, as discussed in Chap. 2. According to this theory, the scattering property of a small particle is determined by two physical parameters, the scattering efficiency Q and the anisotropy value g. The former gives the strength of the scattered field while the latter represents the averaged angle of the scattered field. The Mie scattering theory provides an important physical foundation of the Monte Carlo simulation for optical microscopic imaging through a tissue-like turbid medium consisting of scattering particles such as cells and nucleii [1].

However, these small scattering particles are aggregated because of their biological and chemical functions [17, 18]. An aggregate is a medium with an ensemble of small scattering particles which, due to the physical mechanisms of aggregation, lead to a fractal structure, i.e., a scale-invariant structure [19]. Optical excitations in fractal aggregates are substantially different from those in other media [20]. Both electromagnetic and Monte Carlo analyses have been developed to study the light scattering behavior through a fractal aggregate [21, 22]. However, these methods are not applicable to investigating photon migration through a complex turbid tissue medium that is composed of many aggregates distributed at different sites [22]. To deal with multiple scattering in a turbid medium made up of scattering aggregates, it is necessary to establish a physical model that gives rise to the scattering parameters of one aggregate under plane wave illumination.

The aim of this section is to present a model of light scattering by a spherical aggregate [23]. Using Monte Carlo simulation described in Sect. 3.1, we calculate the angular distribution of photons scattered by a spherical aggregate consisting of small particles that satisfy the Mie scattering theory. Based on this, we derive the dependence of the scattering parameters of the aggregate, the scattering efficiency Q and the anisotropy value g, on its size and the effective mean free path length. This method is termed the effective Mie scattering (EMS) model with which we predict the focal spot in the tissue-like turbid medium made up of scattering aggregates.

3.8.1 Effective Mie Scattering by a Spherical Aggregate

Figure 3.7 shows the schematic diagram of the EMS model for an aggregate. A plane wave of wavelength λ is illuminated on an aggregate for excitation. The scattering aggregate has an imaginary spherical shape of diameter D (radius R) and a fractal dimension of m and consists of spherical scattering particles of diameter d (radius a) [22]. Here m denotes the degree of aggregation and $m = 3$ represents a homogeneously random medium. $2 < m < 3$ gives the fractal degree description of a fractal aggregate in three-dimensional space that is represented by a spherical coordinate system shown in Fig. 3.7.

Due to the structural feature of a fractal medium, the effective mean free path length (l_m) for an aggregate is not simply related to the scattering cross section and the concentration of scattering particles in a homogeneously random medium ($m = 3$) rather than the following expression [23]:

$$l_m = a[4(m-2)/KmQ_{\text{p}}]^{1/(m-2)} \tag{3.35}$$

where K and Q_{p} are, respectively, the volume fraction of the fractal medium and the scattering efficiency of a scattering particle determined by the Mie scattering theory [14].

To understand the scattering properties of a spherical aggregate, one needs three physical parameters, the phase function $\rho(\theta)$, the scattering cross section σ_{s} (or the scattering efficiency Q), and the anisotropy value g. To this end, we first calculate the angular distributions of scattered photons and total photons. Because of the cylindrical symmetry, these two distributions are dependent only on the angle θ, and represented by $N_{\text{s}}(\theta)$ and $N_{\text{s}}(\theta)$. A phase function $\rho(\theta)$ specifies the angular dependence of the photons scattered into a unit solid angle dΩ oriented at a scattering angle θ relative to the original light trajectory. Therefore, the normalized phase function $\rho(\theta)$ of an aggregate can be calculated by

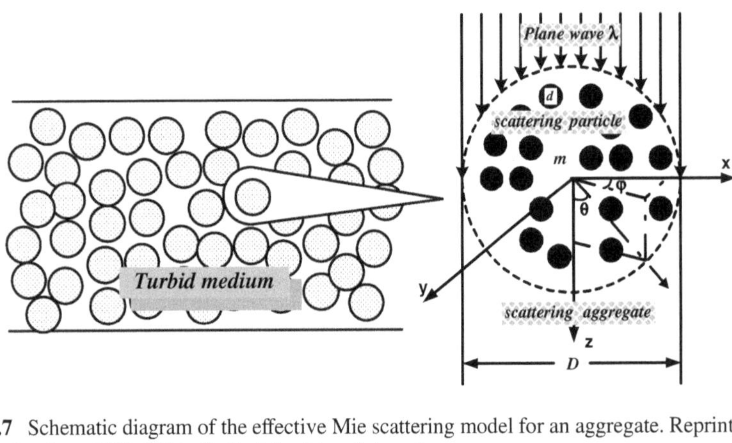

Fig. 3.7 Schematic diagram of the effective Mie scattering model for an aggregate. Reprinted with permission from [23], 2004, Optical Society of America

$$\rho(\theta) = (N_s(\theta)/\sin\theta)/\int_0^{2\pi} d\varphi \int_0^{\pi} (N_s(\theta)/\sin\theta)d\theta \quad (3.36)$$

The scattering cross section σ_s is the ratio of the energy scattered at θ, W_s, to the incident irradiance I_i, i.e., $\sigma_s = W_s/I_i$, while the scattering efficiency Q is the ratio of the scattering cross section of an aggregate to the geometrical cross section σ_g, i.e., $Q = \sigma_s/\sigma_g$. For a spherical aggregate of radius R, the scattering efficiency Q_a of a spherical aggregate can be obtained through

$$Q_a = \int N_s(\theta)d\theta / \int N_t(\theta)d\theta. \quad (3.37)$$

The average cosine of the scattering angle θ of the scattered light is called the anisotropy value g and can be calculated by the use of the phase function $\rho(\theta)$. In our case, the anisotropy value g_a of an aggregate is given by

$$g_a = \int N_s(\theta)\cos\theta\, d\theta / \int N_s(\theta)d\theta. \quad (3.38)$$

3.8.2 Numerical Results

The scattering cross section σ_p and the anisotropy g_p value of a scattering particle within an aggregate can be determined by the Mie scattering theory and the effective mean free path length l_m of an aggregate can be calculated using (3.35). Therefore, the three physical scattering parameters of a spherical aggregate can be calculated using the Monte Carlo simulation method in Sects. 3.1 and 3.5 for a uniform turbid medium [23]. The parameters used in this Monte Carlo simulation are as follows. 10^7 illumination photons are used to ensure the accuracy of simulation results. It is assumed that the excitation wavelength λ is 0.4 μm. The diameter of the aggregate D is 5 μm and the diameter of the scattering particles which constitute the aggregate with a fractal dimension of m = 2.5 is d = 0.2 μm. The scattering particles have no absorption and are suspended in air. The refractive index n of the scattering particles is 1.59.

The phase function $\rho(\theta)$ of the EMS model for a fractal aggregate is demonstrated in Fig. 3.8 for two cases of l_m = 1.0 and l_m = 5.0. As expected, when the effective mean free path length l_m becomes larger (the dashed curves in Fig. 3.8a, b), the angular distribution of the phase function becomes narrower because the longer l_m means the less effective scattering events through the aggregate.

For comparison, the phase function $\rho(\theta)$ from the conventional Mie scattering theory for a solid spherical particle of the size same as the aggregate is also depicted in Fig. 3.8. A comparison of Mie scattering and EMS shows that Mie scattering

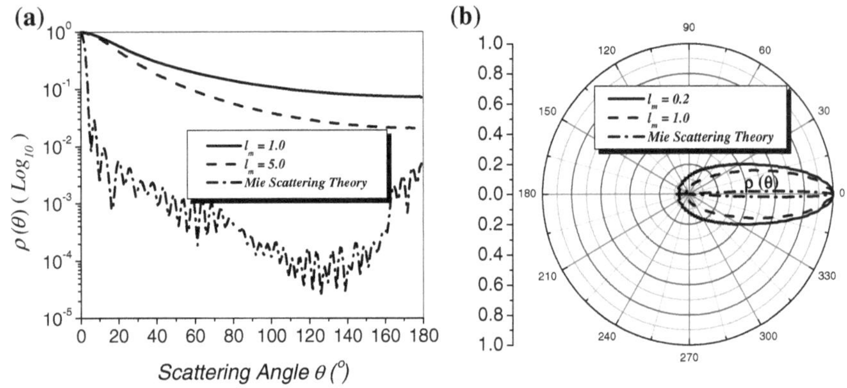

Fig. 3.8 Phase functions $\rho(\theta)$ of an aggregate based on the effective Mie scattering model ($l_m = 1.0$ μm and $l_m = 5.0$ μm) and the phase function $\rho(\theta)$ based on Mie scattering: **a** logarithmic scale; **b** polar coordinate system. Reprinted with permission from [23], 2004, Optical Society of America

leads to a modulating dependence of the phase function (Fig. 3.8a), which does not exist in EMS. This difference results from the different physical mechanisms involved in Mie scattering and EMS. In the former case, the phase function is derived by the diffraction theory and thus the superposition of the field amplitude gives rise to the interference (modulating) feature. However, in the EMS model, the phase function is given by the superposition of photons (see 3.36). Another feature of the comparison is that the solid particle results in a stronger forward scattering characteristic. This feature is caused by the fact that photons are multiple scattered within the aggregate, which gives rise to a broader angular distribution and stronger backscattering (Fig. 3.8b).

The dependence of the scattering efficiency Q_a and the anisotropy value g_a of an aggregate on the normalized aggregate size D/λ is demonstrated in Fig. 3.9a. It can be seen that as D/λ becomes larger, the scattering efficiency Q_a and the anisotropy value g_a become large and small, respectively. This behavior is different from that predicted from the Mie scattering theory (see the dashed curves in Fig. 3.9a) but understood because the number of the scattering events increases as the aggregate size becomes large. In addition, the dependence of the scattering efficiency Q_a and the anisotropy value g_a in EMS does not show the modulating nature, which is consistent with the superposition principle of photons shown in Fig. 3.8.

The effect of the normalized mean free path length l_m/D on the scattering efficiency Q_a and the anisotropy value g_a is demonstrated in Fig. 3.9b). As expected, when l_m/D increase, i.e., when either l_m increases or D decreases, the scattering efficiency Q_a and the anisotropy value g_a decreases and increases, respectively, because of the increased number of the scattering events within the aggregate.

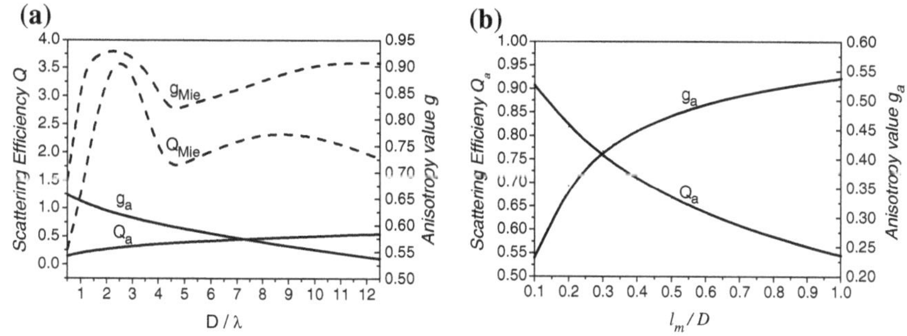

Fig. 3.9 Scattering efficiency Q_a (*left side*) and anisotropy g_a value (*right side*) of an aggregate as a function of its size D/λ (**a**) and mean free path length l_m/D. Reprinted with permission from [23], 2004, Optical Society of America

3.9 Effect of Multi-photon Excitation

The theoretical foundation of two-photon excitation was first established in 1931 by Goppert-Mayer in her doctoral dissertation [24]. Two-photon (2p) excitation arises from the simultaneous absorption of two photons in a single quantized event. It is the most commonly used nonlinear event of multi-photon excitation and was first used for laser scanning optical microscopy in 1990 [25]. The work of Denk et al. [25] launched a new revolution in optical microscopy introduced by two-photon excitation. They successfully demonstrated the 2p fluorescence microscopic imaging of living cells and other microscopic objects. Readers interested in the historical development of two-photon microscopy can refer to the Milestone book by Masters [26].

The basic concept of two-photon excited fluorescence is summarized in Fig. 3.10. Two long wavelength (usually infrared) photons are simultaneously absorbed by the molecules of fluorophores and these two photons cooperate to raise the energy level of one electron in the fluorescent probe from its ground state (S_0) to an excited state. The excited electron subsequently decays back to its ground state and in so doing loses energy in the form of an emitted photon that has a wavelength longer than the wavelength corresponding to the sum of the energy of the two absorbed photons from the light source. This phenomenon is known as the Stokes shift.

The advantages of using two-photon excitation rather than single-photon excitation microscopy arise from the basic physical principle that the absorption depends on the square of the excitation intensity, since the simultaneous absorption of two photons can rarely happen in the out-of-focus region. As a result, the photobleaching and photodamage outside the focal region is greatly minimized and this method thus inherently provides a three-dimensional imaging ability even without the existence of the pinhole that is used in confocal microscopy. Moreover, the use of the longer excitation wavelength for 2p excitation suffers less scattering

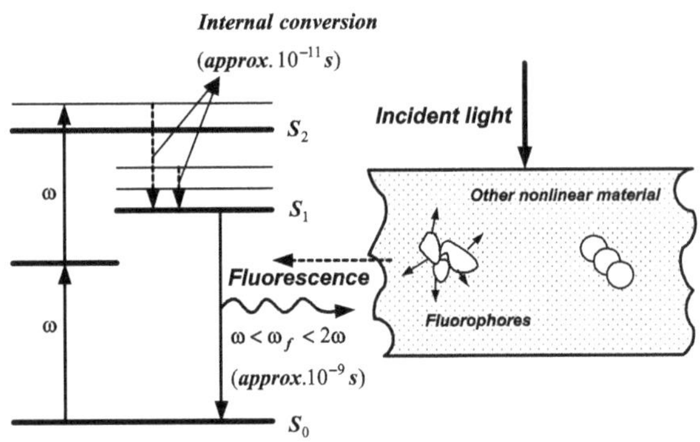

Fig. 3.10 Schematic diagram of two-photon excitation

in a turbid medium, especially in biological tissues, and leads to a deep penetration of imaging. Therefore, 2p fluorescence microscopic imaging, since its first demonstration, has been widely taken as a useful tool for biomedical studies. In a similar justification, one can use three photons to excite biological samples although the excitation efficiency in this case is reduced significantly.

Due to the physical difference in the light interaction with specimen under multi-photon excitation, it is understandable that the effect of multi-photon excitation on imaging formation is accordingly different from that of single photon excitation. Accordingly, the implementation of the Monte Carlo simulation for multi-photon fluorescence microscopy needs some modifications.

Single-photon (1p) fluorescence excitation is a linear process where the excited fluorescence intensity is proportional to the incident intensity

$$I_{1p} = \alpha_{1p} I_{ex} \tag{3.39}$$

where α_{1p} is the 1p fluorescence coefficient, and I_{ex} and I_{1p} are the intensity of the incident and fluorescence signals, respectively. However, multi-photon fluorescence such as 2p or three-photon (3p) fluorescence is a nonlinear process in which case the excited fluorescence intensity has a quadratic or cubic response to the excitation light intensity, respectively:

$$I_{2p} = \alpha_{2p} I_{ex}^2, \quad I_{3p} = \alpha_{3p} I_{ex}^2 \tag{3.40}$$

where α_{2p} and α_{3p} are 2p and 3p fluorescence coefficients, respectively.

Due to the nonlinear process, the implementation of the Monte Carlo simulation under multi-photon excitation (2p or 3p) is not as straightforward as that under 1p excitation. The execution of the Monte Carlo simulation for multi-photon excitation is divided into two stages.

In the first stage, the intensity distribution of the excitation light at the focal plane $I_{ex}(r)$ (where r is the radial distance from the focus) is calculated using the Monte Carlo simulation method introduced in Sect. 3.1 and stored in a database. In the second stage, the same number of photons as those produced in the first stage is

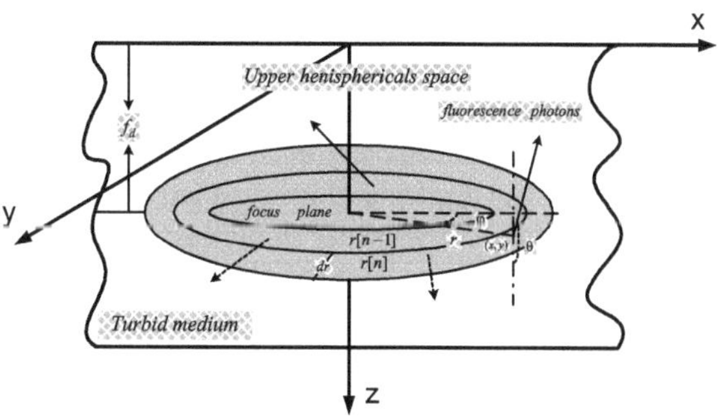

Fig. 3.11 Schematic diagram of the implementation of Monte Carlo simulation under multi-photon excitation

launched in simulation. This time, for each excitation photon, a fluorescence photon is supposed to be produced by assuming $\alpha_n = 1$.

The simulation of the fluorescence photons then starts at the focal plane as shown in Fig. 3.11. The release position of a fluorescence photon is the same as that of its excitation photon $r(x, y)$. However, each fluorescence photon generated at the distance $r(x, y)$ has the weight factor $w(r)$ which is determined according to the weighting functions under 2p and 3p excitation, respectively,

$$w_n(r) = I_{\text{ex}}^{(n-1)}(r) \tag{3.41}$$

where $n = 2$ and 3 corresponding to 2p and 3p excitation, respectively.

Meanwhile, each fluorescence photon is emitted at a random direction centered about the emitting point in the focus plane (x, y). The initial emission directional cosines of a fluorescence photon within the upper hemispherical space can be decided by

$$\begin{cases} u_x = 2.0 * (\xi_x - 0.5) \\ u_y = 2.0 * (\xi_y - 0.5) \\ u_z = -\sqrt{1.0 - (u_x^2 + u_y^2)} \end{cases}, \tag{3.42}$$

where ξ_x and ξ_y are the (0–1) randomly distributed numbers.

3.10 Effect of Coherence

Coherent imaging methods such as optical coherent tomography (OCT) [27] and second harmonic generation (SHG) [28] imaging have recently attracted a great interest since they may provide high resolution images, a deep penetration depth and the special information of a sample.

In the Monte Carlo simulation methods introduced in Sects. 3.1–3.9, light is treated as a bunch of energy packets, i.e., photons, and its wave property is omitted. However, to deal with a coherent process, the phase information from a wave is indispensable. Therefore, in this section, a new Monte Calro model called the coherence Monte Carlo (CMC) simulation is briefly introduced [29]. Based on it, wave phenomena such as coherence and polarization can be studied.

Consider an isotropically distributed spherical particle that is illuminated by a plane wave. The direction of propagation of the incident light defines the z axis, the forward direction. The center of the particle is chosen as the origin of a Cartesian coordinate system (x, y, z) with unit vectors $(\widehat{e}_x, \widehat{e}_y, \widehat{e}_z)$. The scattering direction $\widehat{e}_r$ or $\widehat{e}_k$ (k is the propagation direction of the wavelet) and the forward direction $\widehat{e}_z$ define a plane called the scattering plane (Fig. 3.12). $\widehat{e}_{//}$ and $\widehat{e}_{\perp}$ are the orthonormal unit vectors, which are parallel and perpendicular to the scattering plane, respectively, and satisfy the following relationship:

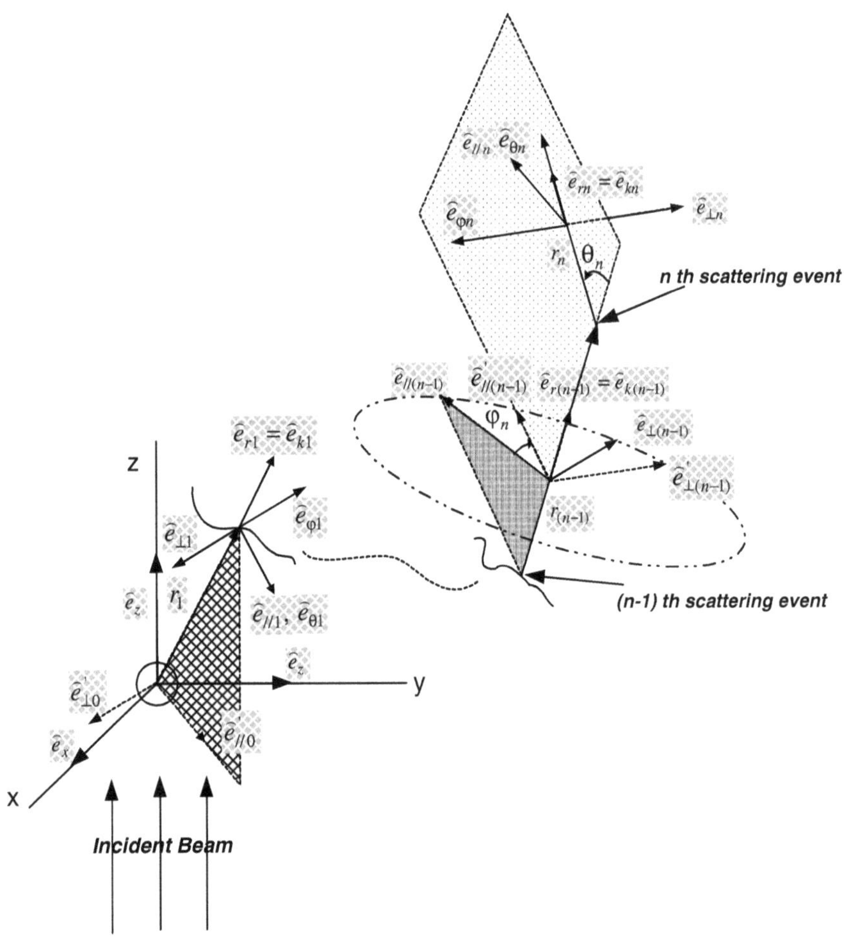

Fig. 3.12 Coordinate systems

$$\widehat{e}_{//} = \widehat{e}_{\theta}, \quad \widehat{e}_{\perp} = -\widehat{e}_{\varphi}, \quad \widehat{e}_{\perp} \times \widehat{e}_{//} = \widehat{e}_{k} = \widehat{e}_{r} \tag{3.43}$$

Accordingly, the electric field E can be expressed by its parallel and perpendicular components $E_{//}$ and $E_{\perp}$. Based on Fig. 3.12, we also have

$$\widehat{e}'_{//0} = \cos\varphi_1 \widehat{e}_x + \sin\varphi_1 \widehat{e}_y, \quad \widehat{e}'_{\perp 0} = \sin\varphi_1 \widehat{e}_x + \cos\varphi_1 \widehat{e}_y \tag{3.44}$$

During the propagation of a wavelet, the phase change is caused by three processes. The first process results from the free transportation between two successive scattering events. The distance (r_n) the wavelet can freely transport between two successive scattering events can be determined by the following equation, which is the same as that in Sect. 3.1:

$$r_n = -\ln(\xi)/\mu_t. \tag{3.45}$$

Therefore, during the transportation of the wavelet, a phase variation is calculated as

$$\Delta\phi_n = kr_n. \tag{3.46}$$

The second process for a phase change arises from the scattering process. The relation between the incident and scattered fields after a single scattering event can be written in a complex matrix form [14]:

$$\begin{pmatrix} E_{//n} \\ E_{\perp n} \end{pmatrix} \sim e^{i\frac{\pi}{2}} \begin{pmatrix} S_2 & 0 \\ 0 & S_1 \end{pmatrix} \begin{pmatrix} E'_{//(n-1)} \\ E'_{\perp(n-1)} \end{pmatrix}. \tag{3.47}$$

Here the scattering matrix components (S_i) due to the scattering from a sphere, in general, is the function of the scattering angle θ, as shown in (2.24) and (2.25).

Finally, the transformation of the electrical field to the scattering plane of the next scattering event leads to ano ther change in phase. From the local scattering coordinate diagram of the photon prior to and after the nth scattering (Fig. 3.12), it is noticed that the $(n-1)$th scattered electric field ($E_{(n-1)}$) and the nth scattered electric field (E_n) are in different local coordinate systems, which are represented by the ($\widehat{e}_{//(n-1)}$, $\widehat{e}_{\perp(n-1)}$, $\widehat{e}_{r(n-1)}$) and ($\widehat{e}_{//n}$, $\widehat{e}_{\perp n}$, $\widehat{e}_{rn}$) coordinate systems, respectively. Between the two systems, a rotation of an azimuthal angle of φ_n happens (Fig. 3.12). In order to understand straightforwardly the relationship of the incident electric field and the next scattered electric field, we project the $(n-1)$th scattered electric field ($E_{(n-1)}$) to be $E'_{(n-1)}$, which is at the same scattering coordinate system ($\widehat{e}_{//n}$, $\widehat{e}_{\perp n}$, $\widehat{e}_{rn}$) as that of (E_n) while a φ_n angle rotation from the coordinate system of $\widehat{e}_{//(n-1)}$, $\widehat{e}_{\perp(n-1)}$, $\widehat{e}_{r(n-1)}$. Based on the coordinate rotation transform, the electric basis vectors of these two systems have the following relationship:

$$\begin{pmatrix} \hat{e}'_{//(n-1)} \\ \hat{e}'_{\perp(n-1)} \end{pmatrix} = \begin{pmatrix} \cos\varphi_n & -\sin\varphi_n \\ \sin\varphi_n & \cos\varphi_n \end{pmatrix} \begin{pmatrix} \hat{e}_{//(n-1)} \\ \hat{e}_{\perp(n-1)} \end{pmatrix}. \tag{3.48}$$

Accordingly,

$$\begin{pmatrix} E'_{//(n-1)} \\ E'_{\perp(n-1)} \end{pmatrix} = \begin{pmatrix} \cos\varphi_n & -\sin\varphi_n \\ \sin\varphi_n & \cos\varphi_n \end{pmatrix} \begin{pmatrix} E_{//(n-1)} \\ E_{\perp(n-1)} \end{pmatrix}. \tag{3.49}$$

Therefore,

$$\begin{pmatrix} E_{//n} \\ E_{\perp n} \end{pmatrix} \sim e^{i\frac{\pi}{2}} \begin{pmatrix} S_2 & 0 \\ 0 & S_1 \end{pmatrix} \begin{pmatrix} \cos\varphi_n & -\sin\varphi_n \\ \sin\varphi_n & \cos\varphi_n \end{pmatrix} \begin{pmatrix} E_{//(n-1)} \\ E_{\perp(n-1)} \end{pmatrix}. \tag{3.50}$$

It is seen that the third process for a phase change comes from the coordinate translation, denoted by the matrix:

$$\begin{pmatrix} \cos\varphi_n & \sin\varphi_n \\ -\sin\varphi_n & \cos\varphi_n \end{pmatrix}.$$

References

1. S.T. Flock, M.S. Patterson, B.C. Wilson, D.R. Wyman, Monte Carlo modeling of light propagating in highly scattering tissues—I: model predictions and comparison with diffusion theory. IEEE Trans. Biomed. Eng. **36**, 1162 (1989)
2. J.M. Schmitt, A. Knuttel, M. Yadlowsky, Confocal microscopy in turbid media. J. Opt. Soc. Am. A **11**, 2226 (1994)
3. X. Gan, M. Gu, Modified Monte Carlo simulation of multi-dimensional photon distribution for microscopic imaging. Optik **108**, 129 (1998)
4. M. Born, E. Wolf, *Principles of Optics* (Pergamon, New York, 1980)
5. S.T. Flock, B.C. Wilson, M.S. Patterson, Total attenuation coefficients and scattering phase functions of tissue and phantom materials at 633 nm. Med. Phys. **14**, 835 (1987)
6. S. Schilders, Microscopic Imaging in Turbid Media, Ph.D. thesis, Victoria University (1999)
7. X. Gan, S. Schilders, M. Gu, Image enhancement through turbid media under a microscope using polarization gating methods. J. Opt. Soc. Am. A **16**, 2177 (1999)
8. X. Gan, S. Schilders, M. Gu, Combination of annular aperture and polarisation gating methods for efficient microscopic imaging through a turbid medium: theoretical analysis. Microsc. Microanal. **3**, 495 (1997)
9. M. Gu, *Advanced Optical Imaging Theory* (Springer, Heidelberg, 2000)
10. S.L. Jacques, L.H. Wang, Monte Carlo modeling of light transport in tissues, in *Optical Thermal Response of Laser Irradiated Tissue*, ed. by A.J. Welch, M.J.C. van Gemert (Plenum Press, New York, 1995), pp. 73–100
11. L.H. Wang, S.L. Jacques, L.Q. Zheng, MCML—Monte Carlo modeling of photon transport in multi-layered tissues. Comput. Methods Programs Biomed. **47**, 131 (1995)

12. A.G. Loewy, P. Siekevitz, *Cell Structure and Function* (A Holt International, New York, 1971)
13. W. Ganong, *Review of Medical Physiology*, 16th edn. (Appleton and Lange, Norwalk, 1993)
14. C.F. Bohern, D.R. Huffman, *Absorption and Scattering of Light by Small Particles* (Wiley, New York, 1983)
15. X. Deng, X. Gan, M. Gu, Monte-Carlo simulation of multi-photon fluorescence microscopy imaging through inhomogeneous tissue-like turbid media, J. Biomedical Opt. **8**, 400 (2003)
16. A. Ishimaru, *Wave Propagation and Scattering in Random Media* (Academic, New York, 1978)
17. A. Wax, C.H. Yang, V. Backman, K. Badizadegan, C.W. Boone, R.R. Dasari, M.S. Feld, Cellular organization and substructure measured using angle-resolved low-coherence interferometry. Biophy. J. **82**, 2256 (2002)
18. A. Wax, C. Yang, M. Müller, R. Nines, C.W. Boone, V.E. Steele, G.D. Stoner, R.R. Dasari, M.S. Feld, In situ detection of neoplastic transformation and chemopreventive effects in rat esophagus epithelium using angle-resolved low-coherence interferometry. Cancer Res. **63**, 3556 (2003)
19. R. Jullien, R. Botet, *Aggregation and Fractal Aggregates* (World Scientific, Singapore, 1987)
20. A. Dogariu, J. Uozumi, T. Asakura, Enhancement of the backscattered intensity from fractal aggregates. Waves Random Media **2**, 259 (1992)
21. Y.L. Xu, Electromagnetic scattering by an aggregate of spheres. Appl. Opt. **34**, 4573 (1995)
22. K. Ishii, T. Iwai, J. Uozumi, T. Asakura, Optical free-path-length distribution in a fractal aggregate and its effect on enhanced backscattering. Appl. Opt. **37**, 5014 (1998)
23. X. Deng, X. Gan, M. Gu, Effective Mie scattering of a spherical aggregate and its application in turbid media. Appl. Opt. **43**, 2925 (2004)
24. M. Göppert-Mayer, Über Elementarakte mit zwei quantensprüngen. Ann. Phys. Lpz **9**, 273 (1931)
25. W.J. Denk, J.H. Strickler, W.W. Webb, Two-photon laser scanning fluorescence microscopy. Science **248**, 73 (1990)
26. B. Masters, *Multiphoton Excitation Microscopy* (SPIE, Bellingham, 2003)
27. G.J. Tearney, M.E. Brezinski, B.E. Bouma, S.A. Boppart, C. Pitris, J.F. Southern, J.G. Fujimoto, In vivo endoscopic optical biopsy with optical coherence tomography. Science **276**, 2037 (1997)
28. P.J. Campagnola, L.M. Loew, Second harmonic imaging microscopy for visualizing biomolecular arrays in cells, tissues and organisms. Nat. Biotech. **21**, 1356 (2003)
29. L. Qiang, X. Gan, Q. Luo, Monte Carlo modeling of optical coherence tomography imaging through turbid media. Appl. Opt. **43**, 1628 (2004)

Chapter 4
Effective Point Spread Function

Abstract Image modeling based on the Monte Carlo method is time consuming because a large number of incident photons are required to ensure a required accuracy. The requiring of computational time increases significantly when images of complicated objects embedded in a turbid medium are modeled. To address this problem, the concept of the effective point spread function (EPSF) has been introduced in this chapter for imaging through turbid media [1]. Section 4.1 provides the detailed conceptual development of the EPSF and numerical simulations of image formation of one-dimensional and two-dimensional objects are detailed in Sect. 4.2.

4.1 Concept of Effective Point Spread Function

If there is no turbid medium, a point spread function (PSF) is the image of a point object and is a measure of image blurring through an imaging system. It determines the resolution of an imaging system and therefore is independent of the property of an object. As a result, under Born's approximation, the image of an object can be obtained from the convolution of a PSF with an object function [1–3]. It should be pointed out that defining PSF in a turbid medium is not straightforward. If an object is embedded in a turbid medium, it is difficult to use the concept of a point object because of the existence of scattering particles which are in the range of 0.1–1 μm in diameter. Further, a PSF that includes only the property of a microscope is not adequate because the multiple scattering effects may severely distort the image of an embedded object [4–11]. Therefore, the property of a microscope as well as the property of scattering particles should be included in the proposed PSF for a microscope.

Without losing generality, a confocal microscope consisting of a finite-sized pinhole is considered in this chapter. In Fig. 4.1, L_1 and L_2 represent the illumination and detection objectives, respectively. P is a finite-sized pinhole of diameter ν_d, placed in front of the detector D (a conventional microscope can be considered to have a large area detector). An effective point spread function (EPSF) for the

M. Gu et al., *Microscopic Imaging Through Turbid Media*,
Biological and Medical Physics, Biomedical Engineering,
DOI 10.1007/978-3-662-46397-0_4

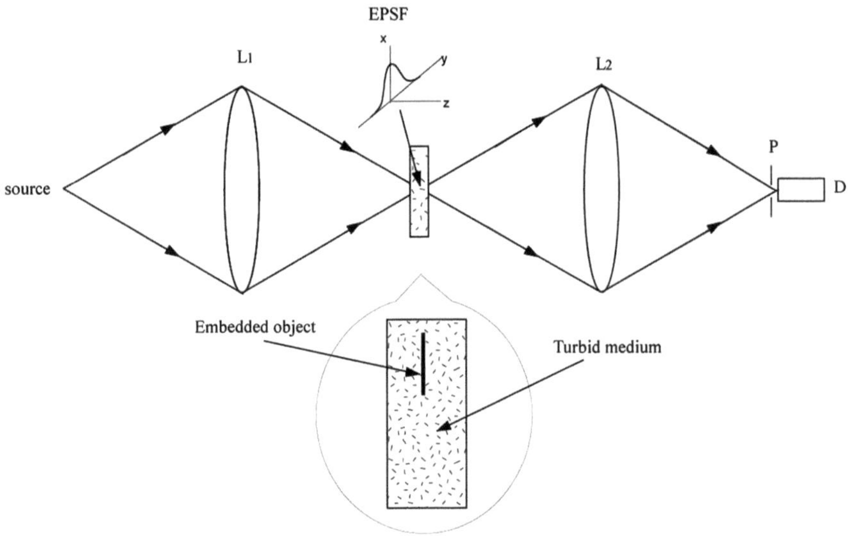

Fig. 4.1 Schematic diagram of a scanning optical microscope. An EPSF in the focal plane is schematically illustrated. A sample includes a turbid medium and an object. Reprinted with permission from [1], 1999, Optical Society of America

microscope can be defined by the distribution of photons in the focal region, which can propagate through a turbid medium, the aperture of a detection objective, and a confocal pinhole. To this end, the coordinates x, y, and z, which are defined with respect to the overlapped focus of the illumination and collection objectives, of each photon passing through a turbid medium is first recorded in the Monte Carlo simulation. Then the simulation carries on to determine whether such a photon can pass through a turbid medium, the aperture of a detection objective, and a confocal pinhole. If a photon reaches the detector D, its contribution is registered in the photon distribution at a point (x, y, z). Otherwise, its contribution is removed. The resulting photon distribution $h(x, y, z)$ is used as the EPSF for the microscope.

It is clear that such an EPSF reflects not only the property of a microscopic imaging system but also the scattering property of a turbid medium. The parameters determining an imaging system include the numerical aperture (NA) of illumination and detection objectives and the size of the confocal pinhole v_d. The parameters which describe a turbid medium are the scattering mean free path length l, the anisotropy value g, and the optical thickness n defined as the sample thickness divided by the scattering mean free path length l. According to the experimental conditions [12–15], we use an example scattering slab which has a thickness (d) of 180 mm and consists of polystyrene beads of diameter 0.48 μm suspended in water. The anisotropy value g of 0.48 μm beads is 0.81 for a He–Ne laser at wavelength 0.633 μm.

4.2 Two Dimensional Case

For a turbid medium of optical thickness $n = 20$, the EPSF as a function of the radial coordinate $r = (x^2 + y^2)^{1/2}$ at the focus ($z = 0$) is shown in Fig. 4.2a for different sizes of the confocal pinhole. The EPSF has been normalized by its value at $r = 0$. Here we have assumed that the focal plane is placed in the middle of the turbid medium. If the focal plane is located at different depths of the turbid medium, the corresponding EPSF is different. It is noticed from Fig. 4.2a that the EPSF becomes narrower when a smaller confocal pinhole is used. The sharp component near $r = 0$ represents the contribution from ballistic photons. Statistically, multiple scattered photons are distributed further away from the center of the detector plane, and thus when a pinhole is used, these photons can be effectively rejected according to their deviation from the path that ballistic photons propagate. Therefore, using a pinhole of smaller size results in a narrower EPSF, which can lead to an image with higher resolution [11]. Figure 4.2b shows the EPSF for different values of the optical thickness n. As expected, the contribution from ballistic photons increases as the optical thickness becomes small.

One may ask if the EPSF derived above satisfies a convolution operation, i.e., if the image intensity $I(x, y)$ of a thin object can be modeled by the convolution of an object function $O(x, y)$ and the EPSF $h(x, y)$:

$$I(x,y) = \iint_{-\infty}^{\infty} h(x,y)O(x-x',y-y')\mathrm{d}x'\mathrm{d}y', \tag{4.1}$$

where $h(x, y)$ is the EPSF in the focal plane. In order to demonstrate the validity of (4.1), and the efficiency and accuracy of the image modeling method in (4.1), we compare images modeled by the convolution algorithm in (4.1) with those modeled by the direct simulation method [12]. The object used for this comparison is a highly absorptive edge embedded in the middle of a turbid medium which has a

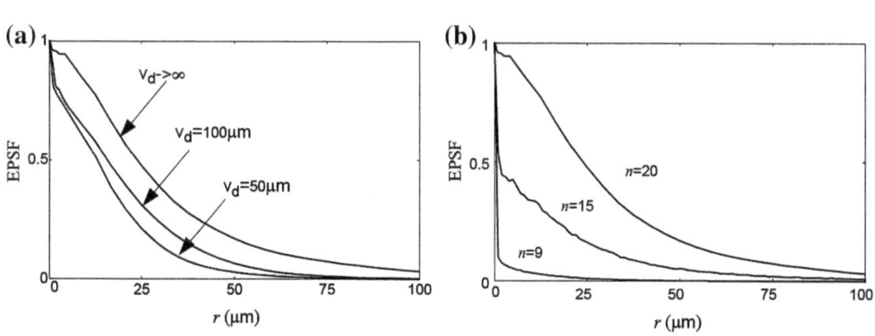

Fig. 4.2 Effective point spread function in the focal plane for **a** different sizes of the confocal pinhole, v_d (NA = 0.25, n = 20) and **b** different values of the optical thickness n (NA = 0.25, v_d = 100 μm). Reprinted with permission from [1], 1999, Optical Society of America

thickness (d) of 180 μm and consists of polystyrene beads of diameter 0.48 μm suspended in water. This sample is scanned in the direction perpendicular to the edge. The illumination wavelength is assumed to be 0.633 μm. From the image intensity of the sharp absorption edge, the transverse resolution is defined as the distance between the 90 and 10 % intensity points.

A comparison of the transverse resolution and computational time in the convolution method with those in the direct simulation method [12] is shown in Table 4.1. The computational time is recorded on a Pentium II 233 MHz PC for a given number of 5,000,000 incident photons for each image pixel. It can be seen from Table 4.1 that the difference in the resolution between the two methods is within 2–3 % for the optical thickness up to 20. This result confirms that the proposed EPSF satisfies the convolution relation in (4.1). The image modeling method which uses the convolution operation in (4.1) is much more efficient in saving computational time than the direct simulation method. For example, for an edge object (Table 4.1), the new method is approximately 15 times faster than the direct simulation time. The computational time in the convolution method mainly arises from the calculation of the EPSF. When the number of incident photons is increased, the computational time of the EPSF increases but the computational time of the convolution operation is not affected.

The advantage of saving computational time in the convolution method is more significant when a two-dimensional (2D) object is modeled. Now we replace the thin edge by a thin circular disk of 20 μm in diameter. According to the computational time summarized in Table 4.1, the estimated computational time for an image of the disk, consisting of 100 × 100 pixels, would be approximately 250 days for $n = 12$ on a Pentium II 233 MHz PC if the direct simulation method were used. However, it takes only 9 h when the convolution method is used. The simulated images of the thin circular disk embedded in turbid media of different optical thicknesses n is shown in Fig. 4.3. It is noticed that when the optical thickness increases, the image becomes blurred as expected. The blurring of the image is caused by the contribution from multiple scattered photons in a turbid medium. This feature can be found from Fig. 4.3 which shows that the image is actually

Table 4.1 Comparison of image resolution and computational time in the convolution method with those in the direct simulation method (Reprinted with permission from [1], 1999, Optical Society of America)

Optical thickness (n)	Convolution method		Direct simulation method	
	Transverse resolution (μm)	Computational time (h)	Transverse resolution (μm)	Computational time (h)
4	29	0.6	30	9.5
9	107	1.7	105	28
15	141	3.6	144	48
20	163	4.9	167	69

The numerical aperture of an objective is 0.25 and the pinhole size is 100 μm

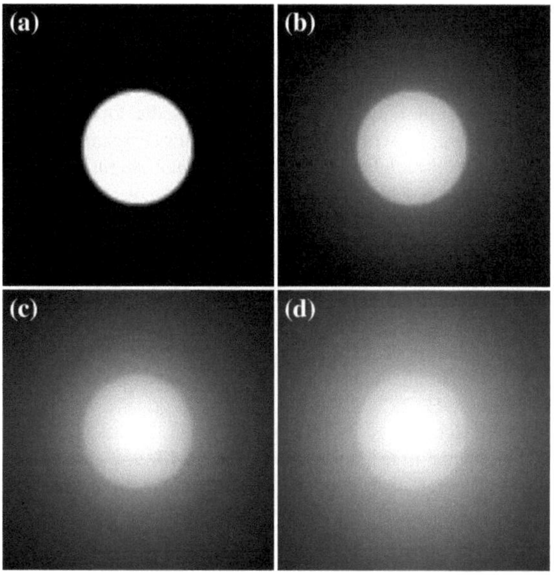

Fig. 4.3 Images of a thin circular disk of 20 μm in diameter (NA = 0.25, $v_d \rightarrow \infty$): **a** $n = 0$; **b** $n = 6$; **c** $n = 9$; **d** $n = 12$. Reprinted with permission from [1], 1999, Optical Society of America

constructed by two components. One component is a sharp image contributed by ballistic photons and the other is a blurred image by scattered photons. With increasing optical thickness, the strength of the ballistic light degrades exponentially and all incident photons are statistically scattered. Thus the sharp image created by ballistic photons gradually fades while the blurred image contributed from scattered photons becomes observable.

The convolution relation in (4.1) holds for any 2D thin object. For a thick object embedded in a turbid medium, the three-dimensional (3D) convolution of a 3D EPSF and a thick object holds only under Born's approximation used in Fourier optics [1, 2], which implies that multiple scattering and secondary diffraction within a thick object is negligible. This approximation does not hold if there is a shadowing effect, for example, in the case of multiple objects. Finally, it should be emphasized that the EPSF defined here is different from the microscope responses in the detector plane calculated in the previous studies [7, 8, 11, 12], the latter does not satisfy the convolution operation even for a thin object.

In conclusion, an EPSF, which incorporates not only the property of a microscope but also the scattering property of a turbid medium in which an object is embedded, has been defined. The proposed EPSF satisfies the convolution operation in (4.1) which offers a much more efficient way in image modeling in comparison with the direct simulation method. Because information other than an embedded object is included in an EPSF, image reconstruction of an object embedded in a thick turbid medium can be possible even when the scattered photons become dominant.

References

1. X. Gan, M. Gu, Effective point-spread function for fast image modeling and processing in microscopic imaging through turbid media. Opt. Lett. **24**, 741 (1999)
2. M. Born, E. Wolf, *Principles of Optics* (Pergamon, New York, 1980)
3. M. Gu, *Principles of Three-Dimensional Imaging in Confocal Microscopes* (World Scientific, Singapore, 1996)
4. C.F. Bohern, D.R. Huffman, *Absorption and Scattering of Light by Small Particles* (Wiley, New York, 1983)
5. Y. Hasegawa, Y. Yamada, M. Tamura, Y. Nomura, Monte Carlo simulation of light transmission through living tissues. Appl. Opt. **30**, 4515 (1991)
6. H. Key, E.R. Davis, P.C. Jackson, P.N.T. Wells, Monte Carlo modelling of light propagation in breast tissue. Phys. Med. Biol. **36**, 591 (1991)
7. J.M. Schmitt, A. Knuttel, M. Yadlowsky, Confocal microscopy in turbid media. J. Opt. Soc. Am. A **11**, 2226 (1994)
8. J.M. Schmitt, K. Ben-Letaief, Efficient Monte Carlo simulation of confocal microscopy in biological tissue. J. Opt. Soc. Am. A **13**, 952 (1996)
9. A. Dunn, C. DiMarzio, Efficient computation of time-resolved transfer functions for imaging in turbid media. J. Opt. Soc. Am. A **13**, 65 (1996)
10. X. Gan, M. Gu, Optik **108**, 129 (1998)
11. X. Gan, S.P. Schilders, M. Gu, Image formation in turbid media under a microscope. J. Opt. Soc. Am. A **16**, 2052–2058 (1998)
12. X. Gan, S. Schilders, M. Gu, Combination of annular aperture and polarization gating methods for efficient in microscopic imaging through a turbid medium: theoretical analysis. Microsc. Microanal. **3**, 495–503 (1997)
13. M. Gu, T. Tannous, C.J.R. Sheppard, Effect of an annular pupil on confocal imaging through highly scattering media. Opt. Lett. **21**, 312 (1996)
14. S. Schilders, X. Gan, M. Gu, Image enhancement through turbid media under a microscope by use of polarization gating methods. Appl. Opt. **37**, 4300 (1998)
15. S. Schilders, X. Gan, M. Gu, Microscopic imaging through a turbid medium by use of annular objectives for angle gating. Appl. Opt. **37**, 5320 (1998)

Chapter 5
Angle-Gating Mechanism

Abstract The key element in an optical microscope is an objective. As a result, an object embedded in a turbid medium is illuminated by a focused beam of angles of convergence determined by the numerical aperture of the objective. Consequently, scattered and unscattered photons are mixed in a range of angles when they leave a turbid medium. In this chapter, an angle-gating mechanism for separating scattered and unscattered photons, achieved by the utilization of polarized annular objectives in the illumination and collection beam paths of microscopic imaging systems, are presented. The principle of angle-gating together with the confirmation by Monte Carlo simulation is described in Sect. 5.1. The performance of annular objectives in transmission and reflection optical microscopes is presented in Sects. 5.2 and 5.3, respectively. In Sect. 5.4, a general issue regarding imaging resolution in turbid media is discussed.

5.1 Principle of Angle-Gating

5.1.1 Concept of Angle-Gating

The concept of angle-gating was first proposed by Gu and Gan [1] in simulating the photon distribution in the time domain and the angle domain. In this simulation, based on the method in Sects. 3.1 and 3.3, an ultrashort pulsed beam with a temporal width, $\Delta\tau_0$, of 20 fs and a central wavelength, λ_0, of 700 nm is used. The turbid medium chosen has a cell geometric thickness, d, of 160 μm, a scattering mean free path length, l, of 40 μm, and an anisotropy value, g, of 0.9, for the central wavelength, λ_0. The incident pulsed beam is focused onto the turbid medium via an objective of numerical aperture 0.25.

Figure 5.1 illustrates the angular-temporal photon distribution after the incident pulsed beam has propagated through the turbid medium [2]. The angular resolution in Fig. 5.1 is 0.5° and the temporal resolution is one-tenth of the pulse width. In Fig. 5.1a, the turbid medium is illuminated by an objective with a circular aperture (ε = 0, where ε is defined in this thesis as the radius of the central obstruction

M. Gu et al., *Microscopic Imaging Through Turbid Media*,
Biological and Medical Physics, Biomedical Engineering,
DOI 10.1007/978-3-662-46397-0_5

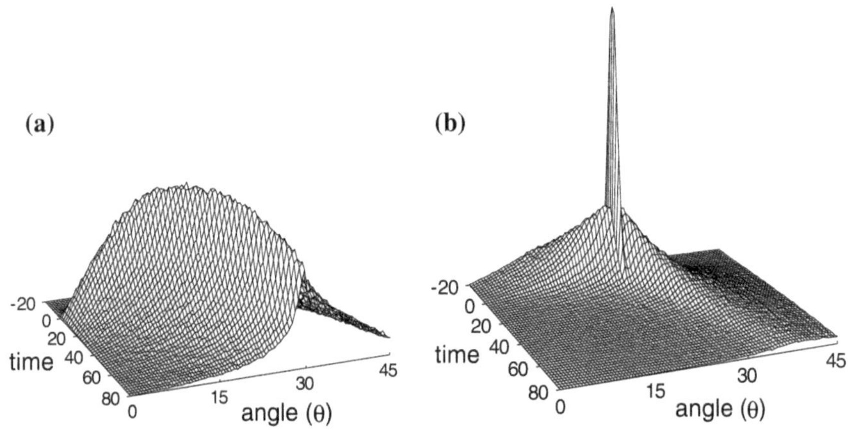

Fig. 5.1 Angular-temporal photon distribution of a 20-fs ultrashort pulsed beam propagating through a turbid slab (l_s = 40 μm, d/l_s = 4). The time t is a local time, so t = 0 corresponds to the arrival time of the unscattered pulse peak. **a** Illumination with a circular objective; **b** illumination with an annular objective. Reprinted with permission from [3], 1997, Cambridge University Press

normalized by the radius of the objective). It is seen that photons propagating at higher angles arrive later than those propagating at lower angles through the objective. It is also noted that there is no clear separation between unscattered and scattered photons in the time domain. This phenomenon demonstrates that time-gating methods are ineffective in suppressing scattered photons. It is also illustrated in Fig. 5.1a that there are no distinguishable unscattered components shown in the angle domain. This phenomenon can also reduce the effectiveness of the angle-gating methods that rely on the directional selectivity to suppress scattered photons. These two phenomena arise from the large range of illumination angles produced by the utilization of a high numerical aperture objective in the imaging process. But high numerical aperture objectives are necessary for high-resolution imaging.

The use of an annular imaging objective to perform an effective angle-gating mechanism [2] is demonstrated in Fig. 5.1b which illustrates the angular-temporal photon distribution under illumination with a thin annular objective (ε = 0.9). An annular lens is simulated in the Monte Carlo model by confining the x–y coordinate of point A (see Fig. 3.1) to a circular ring instead of a circular area. It is noted that under this configuration there is a distinguishable unscattered photon peak in both angular and temporal domains. According to geometric optics, a thin annular aperture confines illumination photons to a small region. Therefore, the exit angles of the unscattered photons can remain in these small regions, while the scattered photons exit in a much broader angle range. Figure 5.1b also shows that once a photon is scattered, the possibility for such a photon returning to the original direction is very small. Therefore, the amount of the scattered photons in the angle region defined by the annular objective is small. This phenomenon is produced under illumination with a thin annular lens provides a physical basis to distinguish the scattered and unscattered photons in the angle domain in a microscopic imaging system.

5.1.2 Angle-Gating in a Microscope

Without losing generality, let us consider a transmission optical microscope. Figure 5.2 shows the concept of angle-gating in a transmission optical microscope, where a darker region includes more ballistic photons. When one employs a pair of circular objectives (O_1 and O_2) in a transmission optical imaging system (Fig. 5.2a), there is no physical way to separate scattered photons in the angle domain of the second (collection) objective O_2. In order to separate these two types of photons, an annular objective (an annular objective can be achieved by placing an annular filter F in the front focal plane of an objective in a 4f system [2]) rather than a circular objective is needed for illumination. The size of an annular objective, ε, is defined as the ratio of the central obstruction radius to the objective aperture radius. $\varepsilon_{\rm in}$ and $\varepsilon_{\rm out}$ are used to define annular objectives in the illumination and collection paths, respectively. It should be noted that the range of the illumination angles for an annular objective is reduced but the numerical aperture of the objective does not change since the maximum aperture of the objective does not change. As expected, with the introduction of an illumination annular objective, it is possible to separate the scattered and unscattered photons from each other since the scattered photons statistically propagate along different angles from the unscattered photons (Fig. 5.2b).

To make use of this unscattered photon peak for high-quality microscopic imaging, one can employ another thin annular filter F_2 to collect the signal originating from the object embedded in turbid media, as depicted in Fig. 5.2c. Although a pair of annular objectives can eliminate scattered photons which do not travel along the same direction of the illumination beam, there are still photons which are

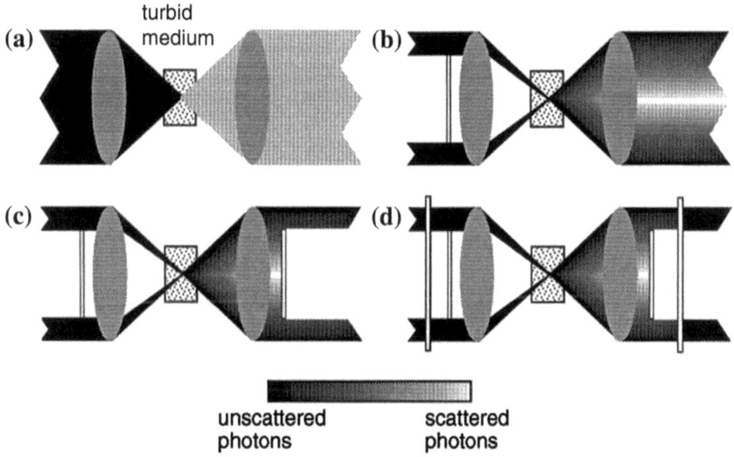

Fig. 5.2 Principle of angle-gating in a transmission microscopic imaging system when **a** a pair of circular objectives, **b** an illumination annular objective and a collection circular objective, **c** a matching pair of annular objectives and **d** a matching pair of annular objectives employing polarization-gating are employed

multiply scattered and finally travel along the same direction as the illumination beam. These scattered photons are superimposed on top of the unscattered photons and can lead to a reduction of image quality (resolution and contrast) even when annular objectives are employed, since multiply scattered photons carry less information of an object embedded in a turbid medium. These scattered photons do not possess the same polarization state as the incident beam because of the scattering process. Therefore to suppress them, one can insert a polarizer LP_1 in front of or after the illumination annular filter (Fig. 5.2d), so that unscattered photons coming from the sample have the same polarization state as the illumination while the scattered photons traveling along the illumination direction are depolarized. The amount of diffusing photons can be appreciably reduced by using another polarizer LP_2, (Fig. 5.2d) which has the same polarization direction as the incident beam and is placed in the back focal plane of the collection objective. As a result of incorporating a polarization detection mechanism, image quality (resolution and contrast) can be further improved.

To confirm the angle-gating mechanism, we can use the Monte Carlo program [3–5] to calculate the intensity distribution over the collection objective aperture O_2 for a continuos wave beam for sample3 and sample5 (see Table 5.1) which was used in the experiments. Illustrated in Fig. 5.3 is the intensity distribution, normalized by the incident intensity, along the radial direction of the collection objective, r, when a circular ($\varepsilon = 0$) or an annular objective ($\varepsilon = 0.9$) is employed for illumination. It can be seen from Fig. 5.3 that an unscattered photon peak appears near the edge of the collection objective aperture only when an annular objective is used for illumination. This result confirms that the thin illumination annular objective confines the unscattered photons to a small angular region on the

Table 5.1 Parameters for the samples used in the transmission optical microscope

Sample number	Scattering medium	Cell thickness, d (μm)	SMFP length, l_s (μm)	Anisotropy value, g	Optical thickness, n (d/l_s)
1	Distilled water	100	–	–	–
2	0.107 μm polysty. microsph. in water	120	416.4[a]	0.076[a]	0.3
3	0.202 μm polysty. microsph. in water	120	70.2[a]	0.31[a]	1.7
4	0.48 μm polysty. microsph. in water	100	19.2[a]	0.81[a]	5.2
5	0.48 μm polysty. microsph. in water	120	19.2[a]	0.81[a]	6.3
6	Semi-skimmed milk	100	30[b]	<0.4[c]	3.3
7	Semi-skimmed milk	300	30[b]	<0.4[c]	10

[a] The parameters are determined by Mie theory (see Table 2.1)
[b] The parameters are experimentally determined [5]
[c] The parameters are determined from the literature [7]

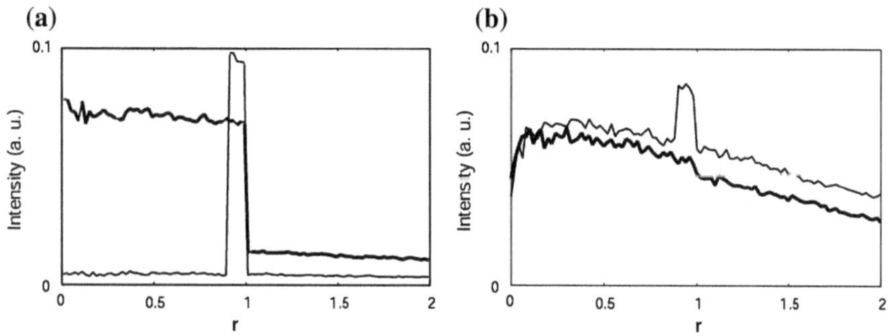

Fig. 5.3 Simulated photon distribution along the radial direction of the collection objective for illumination with a circular objective ($\varepsilon = 0$, *bold*) or an annular objective ($\varepsilon = 0.9$). **a** Sample3 (a turbid medium consisting of 0.202 μm microspheres, see Table 5.1); **b** sample5 (a turbid medium consisting of 0.48 μm microspheres, see Table 5.1). Here ρ is the radial coordinate of the collection objective, normalized by the maximum radius of the objective. Reprinted with permission from [6], 1998, Wiley

collection objective plane. The unscattered photon peak in the case of sample5 (Fig. 5.3b) is not as strong as that for sample3 (Fig. 5.3a) due to the significant difference in the average number of scattering events (see Table 5.1).

5.2 Angle-Gating in Transmission Optical Microscopy

5.2.1 Transmission Optical Microscope and Turbid Samples

The transmission optical microscope shown in Fig. 5.4a is a scanning optical microscope using two polarized annular objectives [5]. A 7 mW Melles Griot (model No. 05-LHP-153) He–Ne laser ($\lambda = 632.8$ nm) is used as the light source. The beam from the laser source is expanded and collimated by an objective O_1 (numerical aperture, NA = 0.25) and a lens L_1 (focal length, $f = 200$ mm, diameter, $D = 25$ mm). The illumination objective O_2 (Zeiss FLUAR 20×/0.75 ∞/0.17) and the collection objective O_3 (Zeiss PLAN-NEOFLUAR 40×/0.75 ∞/0.17) formed a symmetric system around a turbid sample since they have an identical numerical aperture. Scattered and unscattered photons coming from the collection objective are focused via lens L_2 ($f = 200$ mm, $D = 25$ mm) onto a large area detector (1 cm^2) DT which records a time-averaged signal.

Annular objectives can be achieved by placing coaxially opaque circular disc obstructions F_1 and F_2 in the illumination and detection beam paths, respectively. Extreme care should be taken when the focal position of the imaging and collection objectives are overlapped and when the position of the central obstructions in the aperture of the illumination and collection objectives are aligned. The performance of the angle-gating mechanism depends on the successful alignment of these two parameters.

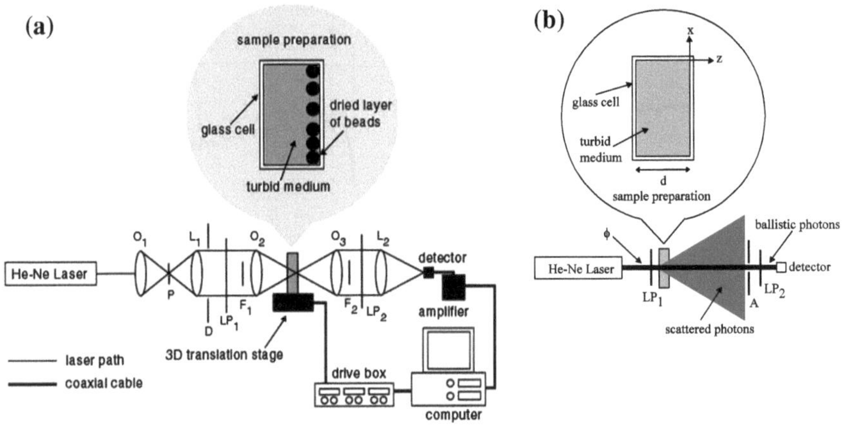

Fig. 5.4 Experimental systems. **a** Transmission scanning optical microscope with angle-gating and polarization-gating mechanisms. *A* aperture; *DT* detector; *F* central obstruction; *L* lens; *LP* polariser; *ND* neutral density filter; *O* objective; *P* pinhole. **b** Optical system to determine the scattering mean free path length, l_s, for semi-skimmed milk. ϕ incident and collection beam diameters; *A* aperture; *LP* polarisers. Reprinted with permission from [5], 1998, Optical Society of America

The polarization-gating mechanism can be implemented by placing a polarizer LP_1 which produced a linearly polarized illumination beam, and a polarizer LP_2 which acts as an analyzer in front of the detector. The degree of polarization, γ, defined in (3.1), of the light passing through a turbid sample can be measured from the signals detected with the analyzer LP_2 parallel (I_p) and perpendicular (I_s) to the direction of LP_1, respectively. The polarization intensity ratio, the ratio of the parallel polarization intensity to the perpendicular polarization intensity, I_p/I_s, (where $I_p/I_s \geq 1$) is also used in the experiments to help demonstrate the contribution made by unscattered and scattered photons. Therefore, a large polarization intensity ratio, I_p/I_s, is desirable since this represents the situation when more unscattered photons are collected than scattered photons.

The details of seven turbid samples are summarized in Table 5.1 [5, 6]. Each sample is placed in an individual glass cell with lateral dimensions of 2 cm × 1 cm. The geometric thickness of the glass cell, d, is varied for the different samples. The optical thickness, n, (i.e., the average number of scattering events experienced by a given photon) is given by, d/l_s for a transmission system, where l_s is the scattering mean free path (SMFP) length, as defined in (2.11). Sample1 consists of distilled water and is used to determine the behavior of the transmission imaging system without scatterers. Sample2, sample3, sample4, and sample5 consist of polystyrene microspheres suspended in water and the scattering parameters of microspheres in these samples are given in Table 2.1. Sample6 and sample7 consist of standard semi-skimmed milk (fat content 3.6 %). Semi-skimmed milk is chosen since it does not include large scatterers (>1 μm), and has been used by Morgan et al. [8] and Wabnitz and Rinneberg [9] in transillumination imaging to simulate tissues which do not include large scatterers (>1 μm). These samples are chosen so that the performance of the angle-gating mechanism can be experimentally investigated in the regimes when scattering is weak and strong.

The diameter of the scattering particles in the milk suspension is assumed to be in the range of 0.1–1.0 μm, and the anisotropy value was assumed from the reference by Jensen [7]. The scattering mean free path length, l_s, for the semi-skimmed milk can be experimentally determined with the experimental arrangement shown in Fig. 5.4b in conjunction with Beers' law shown in (3.1). Rearranging (3.1) yields

$$l_s = \frac{-d}{\ln\left(\frac{N(d)}{N(0)}\right)} \tag{5.1}$$

Therefore as long as the incident power, which is proportional to $N(0)$, and the turbid medium thickness, d, are known, the scattering mean free path length, l_s, can determined by measuring the transmitted unscattered light intensity, which is proportional to $N(d)$, passing through a given turbid medium. The measurement of the transmitted unscattered light intensity is done in the far field region (approximately 1 m away from the turbid sample) to minimize the detection of scattered photons. An aperture in conjunction with a polarization-gating mechanism is employed in front of the detector to assist in minimizing the contribution made by scattered photons, since scattered photons are depolarized and statistically travel along a different direction from unscattered photons.

For evaluating the image quality when polarized annular objectives are used, a cluster layer of 22 μm (in diameter) polystyrene microspheres (standard deviation 2.593 μm, PolyScience Inc.) can be dried onto the inside back surface of the glass cell before it is filled with a turbid medium. The turbid media are placed between the illumination objective and the object, as shown in Fig. 5.4a. Prepared samples are mounted on a three-dimensional piezoelectric translation stage and then placed in the overlapping focal region of the two objectives O_2 and O_3. The translation stages are controlled through a piezoelectric controller. The piezoelectric stages provide 200 μm of traversal with 50 and 100 nm resolution in the x–y and z directions, respectively. The overlapping focus of the two objectives O_2 and O_3 is equally spaced in the center of a turbid sample for the measurements of the degree of polarization, γ. The degree of polarization, γ, for sample1 is measured to be 99.9 % for circular imaging objectives ($\varepsilon = 0$), which shows that depolarization caused by the objectives and other components in this experimental arrangement can be ignored.

5.2.2 *Effect of Annular Illumination and Collection Objectives*

Let us first consider the effect of an annular objective on imaging performance. We take sample4 and sample7 since these two samples can demonstrate the effect of an annular illumination objective when optical scattering is weak and strong. The degree of polarization, γ, with circular illumination and collection ($\varepsilon_{in} = \varepsilon_{out} = 0$) objectives is measured to be 0.574 and 0.118 for sample4 and sample7, respectively. This measurement demonstrates the depolarization of the incident beam due to the scatterers in the turbid media.

Measurements of the degree of polarization, γ, for different central obstruction sizes, ε_{in}, placed in the illumination beam path (F_1) are plotted in Fig. 5.5. Figure 5.5a shows the effect of the central obstruction size of the illumination objective, ε_{in}, on the degree of polarization, γ, when a circular collection objective ($\varepsilon_{out} = 0$), are employed for sample4 and sample7. Figure 5.5b illustrates the polarization intensity ratio, I_p/I_s, for the collected signals under the same conditions. It is seen that as the illumination central obstruction size, ε_{in}, increases from 0 to 0.9, the degree of polarization, γ, decreases from 0.574 (or $I_p/I_s = 3.7$) to 0.532 (or $I_p/I_s = 3.3$) and from 0.118 (or $I_p/I_s = 1.3$) to 0.051 (or $I_p/I_s = 1.1$) for sample4 and sample7, respectively.

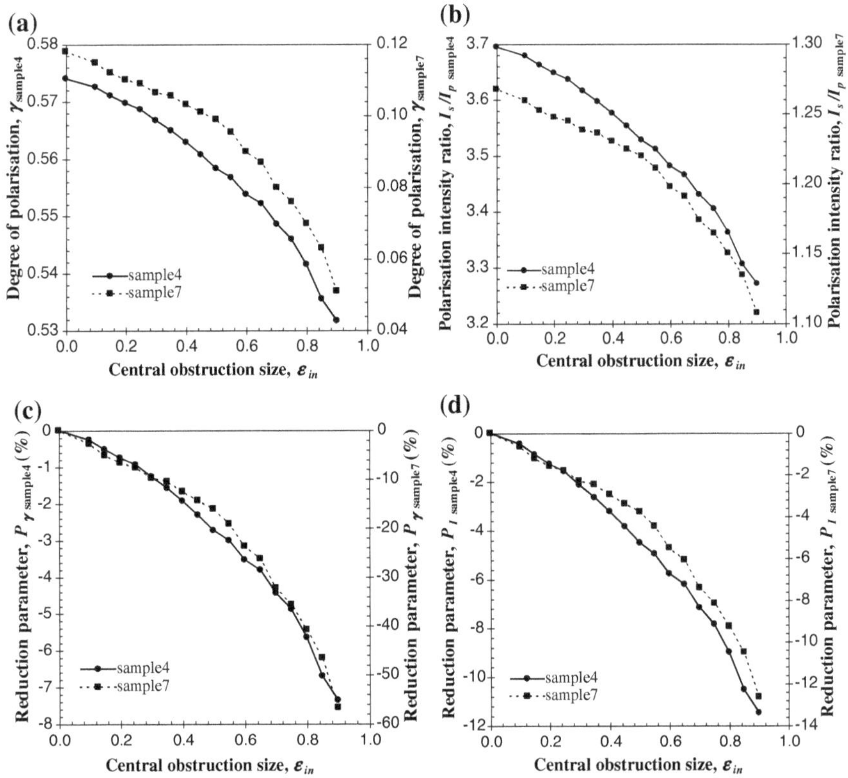

Fig. 5.5 Dependence of the degree of polarization, γ (**a**), the polarization intensity ratio, I_p/I_s (**b**), and the percentage reduction parameters, P_γ (**c**) and P_I (**d**) on the radius of the central obstruction of the illumination objective, ε_{in}. The numerical aperture for objectives O_2 and O_3 is 0.6

To further help demonstrate the performance of the illumination central obstruction size, ε_{in}, on the suppression of scattered photons the following 2 % improvement/reduction parameters are defined

$$P_{\gamma} = \frac{\gamma_{\varepsilon} - \gamma_{\varepsilon=0}}{\gamma_{\varepsilon=0}} \tag{5.2}$$

and

$$P_{\mathrm{I}} = \frac{{}^{I_{\mathrm{p}}}/_{I_{\mathrm{s}}}\Big|_{\varepsilon} - {}^{I_{\mathrm{p}}}/_{I_{\mathrm{s}}}\Big|_{\varepsilon=0}}{{}^{I_{\mathrm{p}}}/_{I_{\mathrm{s}}}\Big|_{\varepsilon=0}}, \tag{5.3}$$

where P_{γ} demonstrates the change in the degree of polarization, γ, and P_{I} demonstrates the change in the polarization intensity ratio, $I_{\mathrm{p}}/I_{\mathrm{s}}$.

The parameters P_{γ} and P_{I} as a function of the illumination central obstruction size, ε_{in}, for sample4 and sample7 are shown in Figs. 5.5c, d, respectively. It is seen that as the illumination central obstruction size, ε_{in}, increases from 0 to 0.9, P_{γ} and P_{I} decrease by approximately 7 and 11 %, respectively, for sample4, while for sample7 the decrease in P_{γ} and P_{I} is approximately 57 and 13 %, respectively. This behavior is because the annular input confines the illumination photons into a high-angle region. Incident photons traveling at higher angles through the turbid samples experience on average more scattering events, leading to a stronger depolarization effect which is indicated by the lower polarization intensity ratio, $I_{\mathrm{p}}/I_{\mathrm{s}}$, when the illumination central obstruction size, ε_{in}, approaches unity (Fig. 5.5b). Since the collection objective collects all the available unscattered and scattered photons within its aperture, scattered photons cannot be further selectively suppressed. Thus the degree of polarization, γ, and the polarization intensity ratio, $I_{\mathrm{p}}/I_{\mathrm{s}}$, inevitably decrease.

Now let us turn to the performance of the central obstruction sizes, ε_{out}, placed in the collection beam path (F_2), on the degree of polarization, γ, for fixed values of the central obstruction size, ε_{in}, placed in the illumination beam path (F_1). Only sample4 is used, since a single turbid sample can demonstrate the effect of the annular collection objective on suppressing scattered photons.

Figure 5.6a shows the dependence of the degree of polarization, γ, on the collection central obstruction size, ε_{out}, (used to suppress the scattered photons) when the central obstruction size, ε_{in}, of the illumination objective is given, while Fig. 5.6b illustrates the polarization intensity ratio, $I_{\mathrm{p}}/I_{\mathrm{s}}$, for the collected signal. The parameters P_{γ} and P_{I} are plotted in Fig. 5.6c, d for the four cases, respectively.

When an annular illumination objective of $\varepsilon_{in} < 0.75$ is employed there is not enough discrimination of the scattered photons even for a collection central obstruction size, ε_{out}, approaching unity. This feature can be further seen from the decreasing values of P_{γ} and P_{I} (i.e., higher contribution of scattered photons) in Fig. 5.6c, d, respectively. It can be clearly seen from Fig. 5.6 that a significant improvement in the degree of polarization, γ, is made only when both central

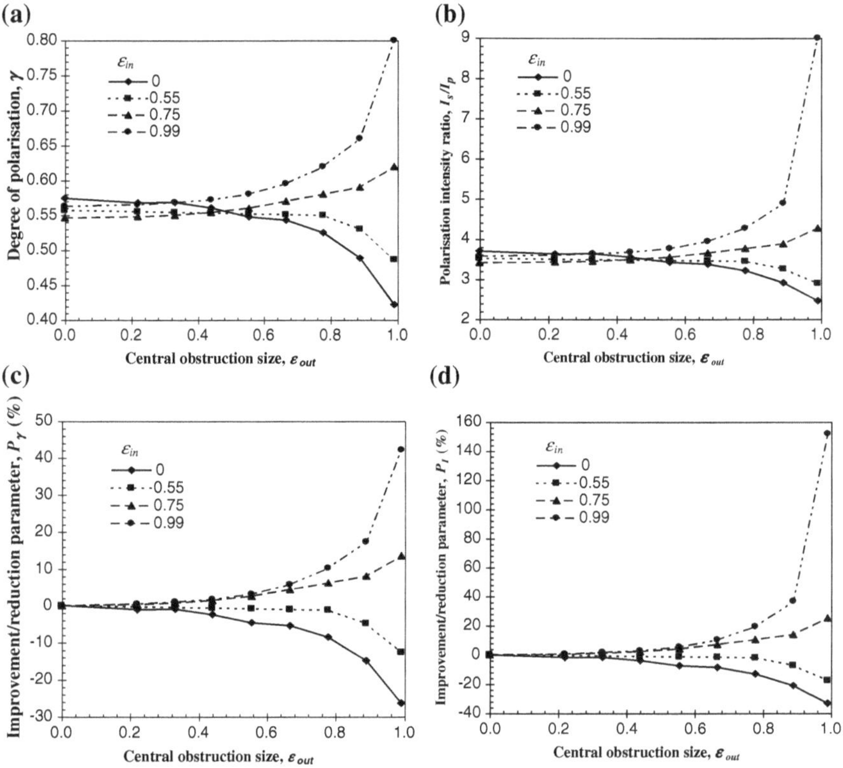

Fig. 5.6 Dependence of the degree of polarization, γ (**a**), the polarization intensity ratio, I_p/I_s (**b**), and the percentage improvement/reduction parameters, P_γ (**c**) and P_I (**d**) on the radius of the central obstruction of the collection objective, ε_{out}, when the central obstruction, ε_{in}, of the illumination objective is given. The numerical aperture for objective O_2 and O_3 is 0.6

obstruction sizes, ε_{in} and ε_{out} are greater than 0.75. For a collection central obstruction size, ε_{out}, equal to 0.99 the improvement in the degree of polarization, γ, is 0.074 (or $P_\gamma \approx 14$ %) and 0.237 (or $P_\gamma \approx 42$ %) when an annular objective with central obstruction size, ε_{in}, equal to 0.75 and 0.99, respectively, is used. The improvement in the polarization intensity ratio, I_p/I_s, for the same conditions is 0.85 (or $P_I \approx 25$ %) and 5.43 (or $P_I \approx 152$ %), respectively. This result suggests that a pair of annular objectives with large central obstructions (i.e., $\varepsilon \rightarrow 1$) should be used to efficiently suppress diffusing photons.

Finally, we examine the performance of two equal central obstruction sizes ($\varepsilon_{in} = \varepsilon_{out} = \varepsilon$) placed in the illumination beam path (F_1) and collection beam path (F_2). In this case sample2, sample3, sample4, and sample7 are used. These four samples are chosen to demonstrate the effect of the angle-gating mechanism on the suppression of scattered photons for turbid media with different scattering characteristics. The dependence of the degree of polarization, γ, on the central obstruction size, ε, for a matching pair of annular objectives and the polarization

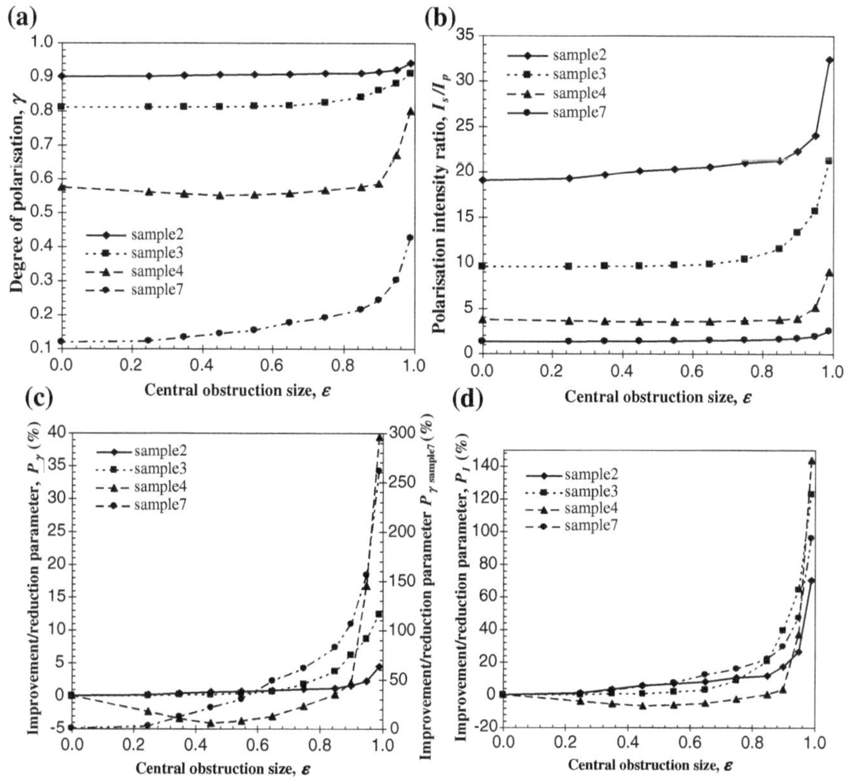

Fig. 5.7 Dependence of the degree of polarization, γ (**a**), the polarization intensity ratio, I_p/I_s (**b**), and the percentage improvement/reduction parameters, P_γ (**c**) and P_I (**d**) on the radius of the central obstruction of a matching pair of annular objectives, ε. The numerical aperture for objectives O_2 and O_3 is 0.6

intensity ratio, I_p/I_s, for the collected signal are shown in Fig. 5.7a, b, respectively. The parameters P_γ and P_I are plotted in Fig. 5.7c, d for the four samples, respectively.

The measured degree of polarization, γ, for the four samples was 0.9 (or $I_p/I_s = 19$), 0.81 (or $I_p/I_s = 9.5$), 0.574 (or $I_p/I_s = 3.7$) and 0.118 (or $I_p/I_s = 1.3$), respectively, when circular objectives ($\varepsilon = 0$) is used. The lower degree of polarization, γ, for sample4 and sample7 results from the fact that the scattering effect is much stronger than that for samples2 and sample3. It is clearly seen from Fig. 5.7a that the degree of polarization, γ, increases appreciably when the central obstruction size, ε, changes from zero to unity. The difference of the degree of polarization, γ, between the matching pair of circular objectives ($\varepsilon = 0$) and the matching pair of annular objectives ($\varepsilon \approx 0.99$) was 0.04 (or $P_\gamma \approx 4$ %), 0.1 (or $P_\gamma \approx 12$ %), 0.226 (or $P_\gamma \approx 39$ %) and 0.308 (or $P_\gamma \approx 261$ %) for sample2, sample3, sample4 and sample7, respectively. The corresponding improvement in the polarization intensity ratio, $I_p/$

I_s is 13.3 (or $P_I \approx 70$ %), 11.7 (or $P_I \approx 123$ %), 5.3 (or $P_I \approx 144$ %) and 1.2 (or $P_I \approx 96$ %) for sample2, sample3, sample4, and sample7, respectively. This result clearly shows that if two thin polarizing annular objectives are used diffusing photons can be efficiently suppressed and that the effect of the polarizing annular objectives on the suppression becomes more significant when the optical thickness, n, becomes larger and depolarization is stronger. The degree of polarization, γ, increases quickly when the central obstruction size, ε, is greater than 0.9, which implies that the improvement in image quality becomes significant in this region. Although an annular objective with the central obstruction size, ε, close to unity allows higher suppression of scattered photons, such a large central obstruction size, ε, reduces the amount of detectable photons appreciably. Therefore, a trade-off value of the central obstruction size, ε, exists between the signal-to-noise ratio of an image and the ability to suppress scattered photons (i.e., resolution achievable).

In Fig. 5.7a sample4 shows that the degree of polarization, γ, decreases first and then increases when the central obstruction size, ε, changes from 0 to 0.9. This phenomenon is caused by the competition between two processes occurring when the central obstruction size, ε, increases. The first process is that increasing the central obstruction size, ε_{in}, of the imaging objective results in more scattering events due to the longer propagation path of scattered photons and leads to a reduction of the degree of polarization, γ, as shown in Fig. 5.5. On the other hand, when a pair of matching annular objectives with a central obstruction size, $\varepsilon > 0.75$ are used, scattered photons can be suppressed and therefore the degree of polarization, γ, increased. The second process becomes dominant when the central obstruction size, ε, approaches unity. As a result, a decreased degree of polarization, γ, occurs for sample4 when the central obstruction size, ε, is not large enough. Once the central obstruction size, ε, is large enough, the degree of polarization, γ, increases appreciably, due to the dominance of the second process. For sample2, sample3, and sample7, it seems that the second process is always dominant. The behavior regarding the effect of the size of the central obstruction, ε, on the degree of polarization, γ, for sample4 is qualitatively confirmed by the theoretical result in Fig. 5.8, which is based on the Monte Carlo method described in Chap. 3 [3].

5.2.3 *Effect of the Numerical Aperture of the Matching Objectives*

In microscopic imaging the goal is to obtain high resolution. To achieve this one needs to employ a high numerical aperture objective. But there is an inherent problem when a high numerical aperture objective is used to image through a turbid medium. Photons propagating at higher angles in such an objective result in longer path lengths that photons traverse, which effectively increases the average number of scattering events experienced by the photons. Therefore, the number of photons maintaining their initial degree of polarization can be reduced.

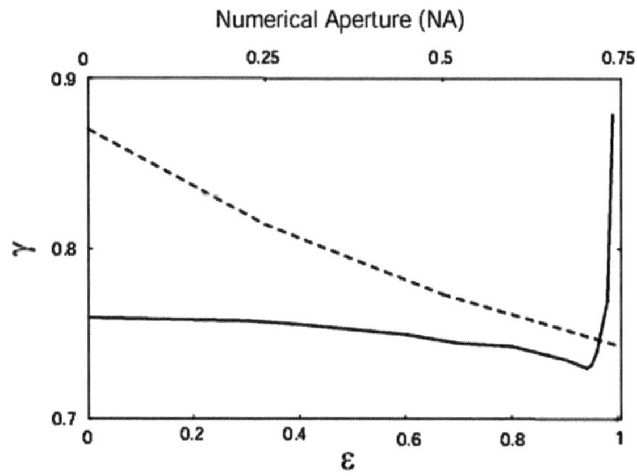

Fig. 5.8 Simulated dependence of the degree of polarization, γ, on the radius of the central annular obstruction for a matching pair of annular objectives with a numerical aperture of 0.6 (*solid curve*) and the numerical aperture of the circular objectives for sample4 in a transmission microscope (*dashed curve*). Reprinted with permission from [5], 1998, Optical Society of America

The effect of reducing the illumination and collection apertures A_1 and A_2 (see Fig. 5.4a) is shown in Fig. 5.9. Reducing the beam diameter of the objective lenses reduces the effective numerical aperture of the objective as well as the angular distribution of illumination and collected photons. Only sample4 and sample7 are considered. The improvement parameters for demonstrating the performance of the numerical aperture (NA) on the suppression of scattered photons are defined as

$$P_\gamma = \frac{\gamma_{\mathrm{NA}} - \gamma_{\mathrm{NA}=0.75}}{\gamma_{\mathrm{NA}=0.75}}, \tag{5.4}$$

and

$$P_\mathrm{I} = \frac{{}^{I_\mathrm{p}}\!/_{I_\mathrm{s}}\Big|_{\mathrm{NA}} - {}^{I_\mathrm{p}}\!/_{I_\mathrm{s}}\Big|_{\mathrm{NA}=0.75}}{{}^{I_\mathrm{p}}\!/_{I_\mathrm{s}}\Big|_{\mathrm{NA}=0.75}}. \tag{5.5}$$

It is clearly seen from Fig. 5.9a that the degree of polarization, γ, increases appreciably for both samples when the numerical aperture is decreased, while Fig. 5.9b illustrates the polarization intensity ratio, $I_\mathrm{p}/I_\mathrm{s}$, for the collected signal. The parameters P_γ and P_I are plotted in Fig. 5.9c, d for the two samples, respectively. The difference in the degree of polarization, γ, between NA = 0.75 and NA = 0.12 is 0.041 (or $P_\gamma \approx 7$ %) and 0.074 (or $P_\gamma \approx 70$ %) for sample4 and sample7, respectively. For the same conditions, the improvement in the polarization intensity ratio, $I_\mathrm{p}/I_\mathrm{s}$ is 0.472 (or $P_\mathrm{I} \approx 13$ %) and 0.202 (or $P_\mathrm{I} \approx 16$ %) for sample4 and sample7, respectively. The improvement in P_γ and P_I is due to the suppression of the higher angle photons by reducing the numerical aperture of the objectives.

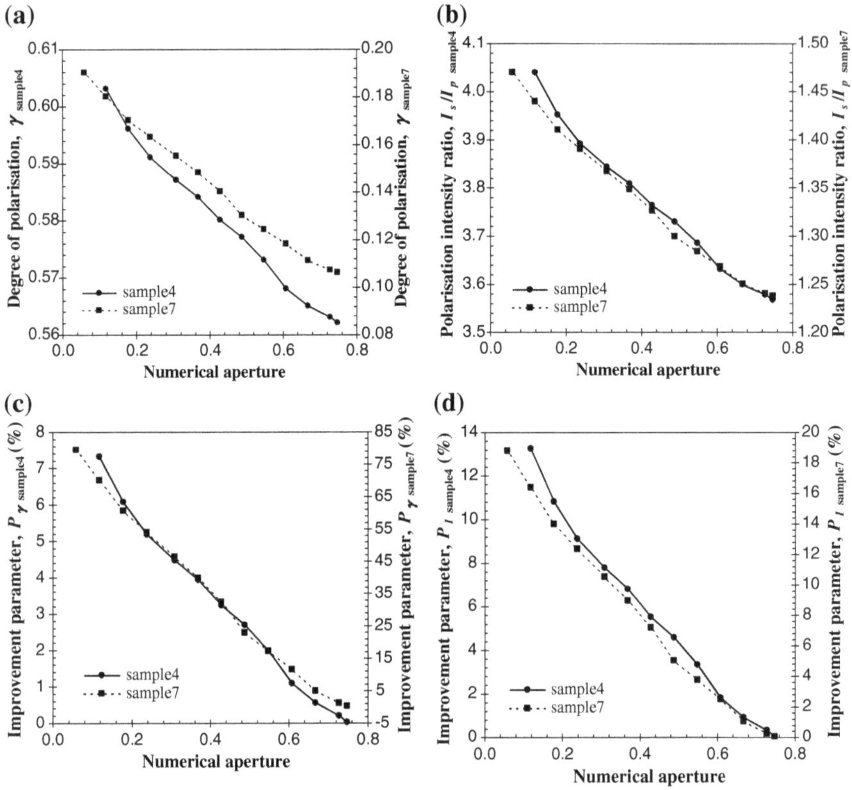

Fig. 5.9 Dependence of the degree of polarization, γ (**a**), the polarization intensity ratio, I_p/I_s (**b**), and the percentage improvement/reduction parameters, P_γ (**c**) and P_I (**d**) on the numerical aperture of the matched illumination and collection objectives

Therefore a trade-off numerical aperture exists between resolution and the ability to efficiently suppress the diffusing photons (i.e., a high degree of polarization, γ). Figure 5.9 confirms the dependence of the degree of polarization, γ, on the numerical aperture which was also observed in sample4.

5.2.4 Imaging with Circular and Annular Objectives

For a better suppression effect as suggested by Fig. 5.9, the numerical aperture of the illumination and collection objectives chosen for all imaging (Figs. 5.10, 5.11, 5.12, 5.13, and 5.14) is 0.25. It should be noted that the detectable signal severely decreases as the central obstruction size, ε, approaches unity. An annular objective with $\varepsilon \approx 0.93$ is therefore chosen to perform all imaging since this annular objective has a good suppression ability and a high signal-to-noise ratio (the detectable light level is significantly above the noise level of the detector used).

The incident power from the laser is attenuated such that the illumination power at the focus for a pair of circular objectives ($\varepsilon = 0$) is equivalent to the power at the focus for a pair of annular objectives ($\varepsilon \approx 0.93$). In this way, a direct comparison of the signal-to-noise ratio (SNR is quantified by the ratio of peak-to-peak signal level to peak-to-peak noise level) can be obtained when circular or annular objectives are employed. As a result, the power at the focus is approximately 110 μW. Polarization-gating is achieved when the polarizer LP_1 and analyzer LP_2 have the same orientation (see Fig. 5.4a). The plane of the overlapped focus for all imaging (unless otherwise stated) is on the surface of the dried bead layer on the turbid medium side. Images are obtained by raster scanning a sample with the aid of a computer.

Figure 5.10 demonstrates the image quality when a dried cluster layer of the 22-μm microspheres embedded within sample1 (see Table 5.1) was imaged with two polarizing circular objectives (Fig. 5.10a, c) and two polarizing annular objectives (Fig. 5.10b, d). Figure 5.10c and d have a reduced field of view of Fig. 5.10a, b to demonstrate the image quality for a single 22-μm microsphere. Since an annular objective has a large depth of focus [10], the image formed with annular objectives appears slightly blurred due to the more contribution from the out-of-focus information (Fig. 5.10b, d) when compared with the images formed with circular objectives (Fig. 5.10a, c). Since the dried microsphere layer is not uniformly dried on the glass surface, the microsphere diameter in the case of circular objectives is dependent on the displacement of the focal position relative to the microsphere surface (Fig. 5.10a). In the case of annular objectives (Fig. 5.10b), the imaged microspheres have approximately the same diameter due to the large depth of focus. All images in Fig. 5.10 are formed by unscattered photons obeying the diffraction theory of light [11]. Therefore, blurring caused by diffraction can become more pronounced as the central obstruction size, ε, for an annular objective is increased in imaging without turbid media. Image contrast in both cases appears black within the shadow of the microsphere, as may be expected from the diffraction by a spherical microsphere.

To demonstrate the effect of turbid media on image quality, sample4 including a cluster layer of 22-μm beads is imaged with circular ($\varepsilon = 0$) and annular ($\varepsilon \approx 0.93$)

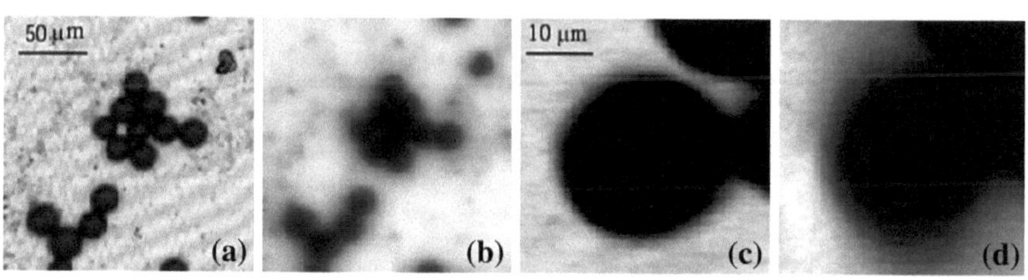

Fig. 5.10 Images of a 22-μm polystyrene microspheres embedded within sample1. **a** and **c** Circular objectives ($\varepsilon = 0$) with parallel polarizers; **b** and **d** annular objectives ($\varepsilon \approx 0.93$) with parallel polarizers. The intensity of the images is normalized to unity. **c** and **d** Are reprinted with permission from [6], 1998, Wiley

objectives. Figure 5.11 shows the images recorded at different focal depths in the sample. It is seen that for circular objectives (Fig. 5.11a, b), there is an inversion of image contrast, which is indicated by the white areas in the center of the microspheres. This inversed contrast can lead to an incorrect interpretation of the information in the image. It is also seen that the image contrast drops dramatically as the focal position moves away from the layer of dried microspheres. In contrast to the case of circular objectives, the images obtained by annular objectives (Fig. 5.11d–f) show the correct contrast and the ability to image the dried microsphere layer even when the focal position is shifted to 40 μm above the surface of the dried microsphere layer. In addition, the pair of annular objectives can enhance the contrast of embedded structures within the turbid media. This feature is indicated by the structure marked by 'A' in Fig. 5.11d. This embedded structure has strong and equal contrast when imaged with annular objectives, whereas when the embedded structure is imaged with circular objectives (Fig. 5.11a) it is difficult to see due to poor contrast.

Figure 5.12 shows the image of a 22-μm microsphere embedded in an optically thicker turbid medium (sample5) under different experimental conditions. In Fig. 5.12a, two circular objectives ($\varepsilon = 0$) are used. Figure 5.12b displays the image when two parallel polarizers are included in the circular objectives ($\varepsilon = 0$). It is seen that the introduction of the polarizers produces a slight improvement in image

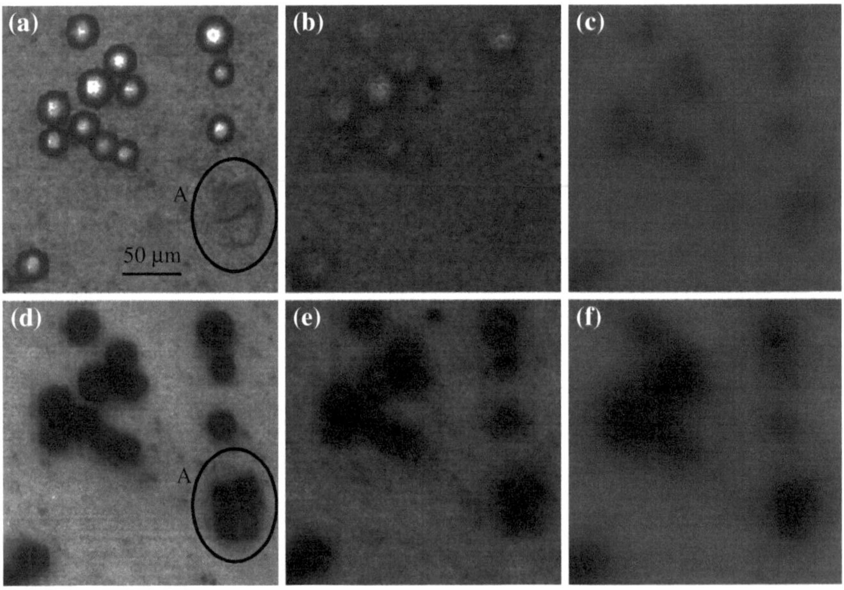

Fig. 5.11 Images of a 22-μm polystyrene microsphere cluster embedded within sample4. **a–c** Circular objectives ($\varepsilon = 0$) with parallel polarization; **d–f** annular objectives ($\varepsilon \approx 0.93$) with parallel polarization. In **a** and **d** the focus is on the dried cluster layer, in **b** and **e** the focus is shifted away from the cluster layer by approximately 20 μm, and in **c** and **f** the focus is shifted away from the dried cluster layer by approximately 40 μm. The intensity of the images is normalized to unity (Reprinted with permission from [5], 1998, Optical Society of America)

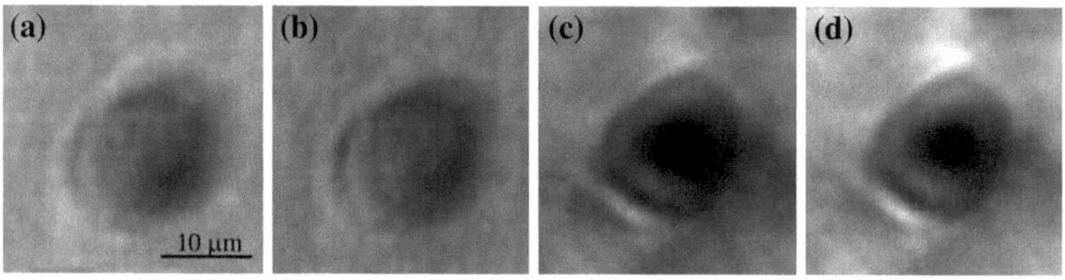

Fig. 5.12 Images of a 22-μm polystyrene microsphere embedded within sample5. **a** Circular objectives (ε = 0) without polarizers; **b** circular objectives (ε = 0) with polarizers; **c** annular objectives ($\varepsilon \approx 0.93$) without polarizers; **d** annular objectives ($\varepsilon \approx 0.93$) with polarizers. The intensity of the images is normalized to unity. Reprinted with permission from [6], 1998, Wiley

detail. Both Fig. 5.12a, b reveal a reversion of image contrast near the center of the microsphere where a dark region should be displayed (see Fig. 5.10c), which as stated earlier may be misleading when the image is interpreted. Figure 5.12c shows the image obtained for two annular objectives ($\varepsilon \approx 0.93$). It can be seen that the images recorded with annular objectives have higher resolution, which is indicated by the smaller microsphere size and reduced smearing. In addition, the images recorded with annular objectives have a correct image contrast in the center of the microsphere, resulting from the effective suppression of scattered photons. A further improvement in the image contrast near the surface of the microsphere can be seen in Fig. 5.12d when the parallel polarizers are included.

Finally, let us demonstrate the performance of polarizing annular objectives on image quality for more realistic samples (i.e., sample6 and sample7). Figure 5.13 shows the images of a dried cluster layer of microspheres embedded in sample6 recorded with circular (Fig. 5.13a) and annular (Fig. 5.13b, c) objectives. In the case of using circular objectives (ε = 0), an inversion of the image contrast is present, which is indicated by the bright central regions. Figure 5.13b, c shows the correct image contrast when annular objectives ($\varepsilon \approx 0.93$) are used. It should be noted that

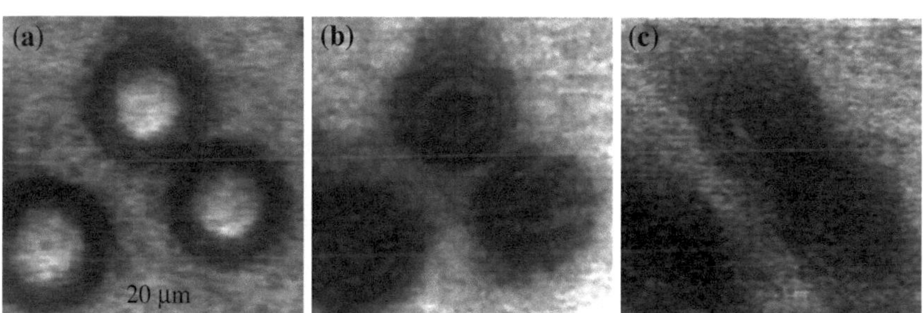

Fig. 5.13 Images of 22-μm polystyrene microspheres embedded in sample6. **a** Circular objectives (ε = 0) with parallel polarization; **b** annular objectives ($\varepsilon \approx 0.93$) with parallel polarization; **c** annular objectives ($\varepsilon \approx 0.93$) with perpendicular polarization. The intensity of the images was normalized to unity (Reprinted with permission from [5], 1998, Optical Society of America)

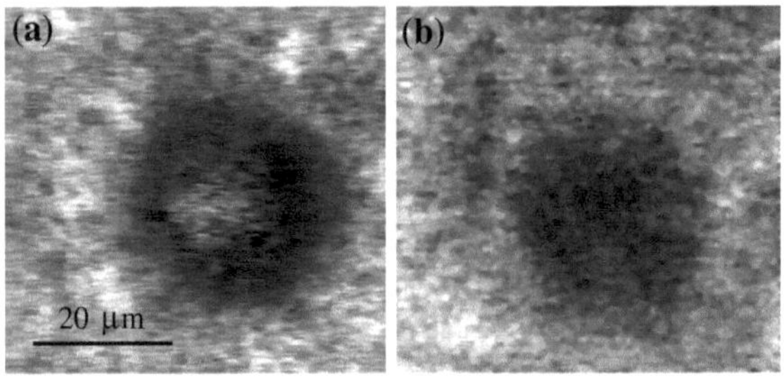

Fig. 5.14 Images of a 22-μm polystyrene microsphere embedded in sample7. **a** Circular objective ($\varepsilon = 0$) with parallel polarization; **b** annular objectives ($\varepsilon \approx 0.93$) with parallel polarization. The intensity of the images is normalized to unity. Reprinted with permission from [5], 1998, Optical Society of America

Fig. 5.13b is obtained with a parallel analyzer LP_2 with respect to the polarizer LP_1, (mainly unscattered photons are detected), while Fig. 5.13c is obtained with a perpendicular analyzer with respect to LP_1 (mainly scattered photons are detected). It is evident that when scattered photons are used to construct an image the resolution drops substantially, which is indicated by the blurring and smearing effects seen in Fig. 5.13c. The full-width-at-half-maximum of the imaged microsphere for the annular (Fig. 5.13b) and circular objectives (Fig. 5.13a) is approximately 28.6 and 28.9 μm, respectively, which indicates that annular objectives can slightly increase resolution, due to the suppression of scattered photons.

To demonstrate the ability of annular objectives to select unscattered light in a thick turbid medium, sample7 including the microsphere cluster layer was imaged with circular (Fig. 5.14a) and annular objectives (Fig. 5.14b). The images recorded with circular objectives ($\varepsilon = 0$) again demonstrate an inversion of image contrast near the center of the microsphere. In addition, severe degradation in image quality is seen, as a result of the increased scattering experienced. Once again, the correct image contrast is seen when the microsphere is imaged with annular objectives ($\varepsilon \approx 0.93$). However, the signal-to-noise ratio of the recorded image (Fig. 5.14b) is degraded due to the reduced number of available unscattered photons. The full-width-at-half-maximum of the imaged microsphere for the annular (Fig. 5.14b) and circular (Fig. 5.14a) objectives is approximately 39.1 and 41.8 μm, respectively. This result further confirms that annular objectives lead to increased resolution in imaging through thick turbid media due to the ability of suppression scattered photons.

5.2.5 Discussion

In Figs. 5.11a, b, 5.13a, and 5.14a, the image contrast is inverted within the shadow of microspheres when circular objectives are used. A possible explanation of this behavior is that scattered photons, which cannot be efficiently suppressed by circular objectives, can reach the shadow area of the microspheres thus contributing to a higher intensity in this region. These scattered photons are considerably suppressed if annular objectives are used, which gives enhanced and correct image contrast. Although annular objectives achieve the correctly enhanced image contrast (Figs. 5.12c, d, 5.13b, c and 5.14b), originating from the angle-gating mechanism outlined in Sect. 4.2, there are a number of issues that need to be discussed regarding their performance as a gating mechanism.

The Monte Carlo method, described in Chap. 3, is based on Mie theory and ignores the diffraction effect. It can be mainly used to describe the behavior of scattered photons in turbid media according to ray optics. In fact, the dominant propagation behavior of unscattered photons in turbid media obeys the prediction by the diffraction theory of light based on Born's approximation [11] which neglects multiple scattering. These two processes, the multiple-scattering of scattered photons and the diffraction by unscattered photons, exist in imaging through turbid media simultaneously. Scattered photons contribute to an image with low resolution whereas unscattered photons give high resolution. When a pair of annular objectives are used to suppress scattered photons, the contribution from unscattered photons is relatively increased with respect to scattered photons when annular objectives are used. As a result, the diffraction effect in the case of imaging in a turbid medium with two annular objectives should be more pronounced, compared with two circular objectives. This property is clearly demonstrated in Fig. 5.13b, c where more diffraction rings are observed. Another diffraction property associated with an annular objective is that an annular objective can transfer more high transverse spatial frequencies than a circular objective [11], which consequently leads to improved transverse resolution as observed in the experiments with annular objectives when the contribution from unscattered photons becomes dominant.

An annular objective possesses a longer focal depth than a circular objective in terms of the diffraction theory [11]. This property is important in achieving angle-gating for a thick object because a longer overlapping focus provided by a pair of annular objectives can efficiently transfer unscattered and less scattered photons from the illumination objective O_2 to the collection objective O_3. This feature is clearly demonstrated in Fig. 5.14, in which circular objectives give a stronger reversion of the image contrast than annular objectives. However, a longer focal depth may be disadvantageous for three-dimensional imaging.

Employing an analyzer with the polarization direction perpendicular to that of the illumination beam actually suppresses unscattered photons. The resultant image (Fig. 5.13c), constructed mainly by scattered photons, shows low contrast and low resolution. This result confirms that scattered photons can build up an image only with low resolution [4]. Figures 5.13b and 5.14b imply that this new method based on polarizing annular gating can give tens-of-micrometer image resolution in this application.

The depolarization of the scattered photons is dependent on the size of scattering particles in turbid media [3]. Under the condition of Fig. 5.13, the depolarization is not so strong that the improvement caused by the polarizers is not significant. However, the image quality can be improved further if the image is constructed by the signal, I_p–I_s, called the differential polarization signal (i.e., the image in Fig. 5.13c is subtracted from the image in Fig. 5.13b). The differential polarization-gating mechanism is the topic in Chap. 6.

5.3 Angle-Gating in a Reflection Optical Microscope

5.3.1 *Reflection Optical Microscope and Turbid Samples*

It has been demonstrated in the previous section that an angle-gating mechanism can be used as an effective tool in suppressing scattered photons and gives the correct image contrast in a transmission scanning microscope in imaging through turbid media. This section demonstrates the implementation of the angle-gating method for suppressing scattered photons by the use of a polarized annular objective in a reflection scanning microscope, which is more applicable for imaging biological tissue in situ (i.e., skin biopsy employing an endoscope).

A reflection scanning optical microscope using polarizing annular objectives is depicted in Fig. 5.15 [12]. A 7 mW Melles Griot (model No. 05-LHP-153) He–Ne laser (λ = 632.8 nm) is used as the light source. The beam from the laser source is expanded and collimated by an objective O_1 (NA = 0.25) and a lens L_1 (focal length, f = 200 mm, diameter, D = 25 mm). In this arrangement, a single objective

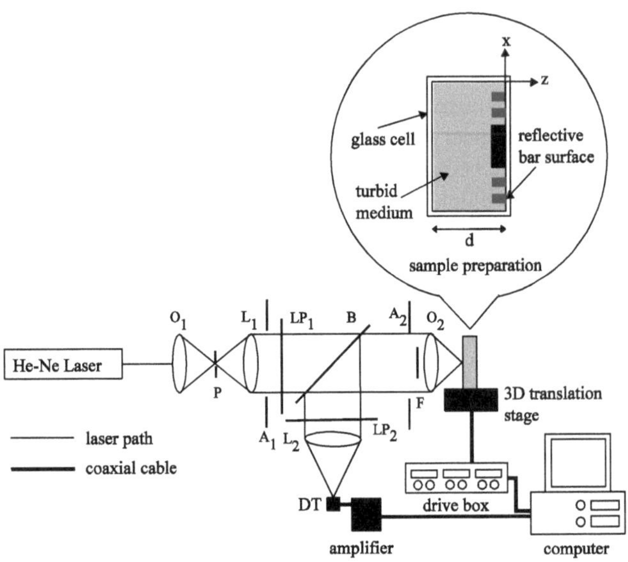

Fig. 5.15 Reflection scanning optical microscope with angle-gating and polarization-gating. *A* aperture; *B* beamsplitter; *DT* detector; *F* central obstruction; *L* lens; *LP* polarizer; *ND* neutral density filter; *O* objectives; *P* pinhole. Reprinted with permission from [12], 1998, Cambridge University Press

O_2 (Zeiss FLUAR™ 20×/0.75 ∞/0.17, hereafter called the imaging objective), instead of two separate objectives as in the transmission optical arrangement (see Fig. 5.4a), is used to illuminate the object embedded within a turbid sample and to collect the diffused and unscattered photons originating from a turbid sample. These collected photons are focused via a lens L_2 ($f = 200$ mm, $D = 25$ mm) onto a large area (1 cm^2) detector DT which records a time-averaged signal.

An annular imaging objective is achieved by placing coaxially an opaque central circular disc obstruction F in the illumination/detection beam path. Like in the case of the transmission system, extreme care should be taken in obtaining the correct alignment of the central obstruction in the aperture of the imaging objective, since the performance of the angle-gating mechanism depends on the alignment of the central obstruction. In order to incorporate the polarization-gating mechanism into the angle-gating mechanism, a polarizer LP_1 is placed such that it produced a linear polarized illumination beam, while the polarizer LP_2 acts as an analyzer in front of the detector. The degree of polarization, γ, (3.1) of light passing through a turbid medium is once again measured from the signal detected with the analyzer LP_2 parallel (I_p) and perpendicular (I_s) to the direction of LP_1, respectively.

Three turbid samples (sample8, sample9, and sample10) are selected and their details are summarized in Table 5.2. Each sample is placed in an individual glass cell with lateral dimensions of 2 cm × 1 cm. The glass cell thickness, d, was 120 μm for the three samples. It should be noted that the optical thickness, n, for a reflection microscope is given by, $2d/l_s$. Sample8 consists of distilled water and is used to determine the behavior of the reflection imaging system without scatterers. Sample9 and sample10 consist of polystyrene microspheres of diameter 0.48 μm suspended in water. It should be noted that, the concentration of polystyrene microspheres for sample9 is diluted by a factor of 2.

As depicted in Fig. 5.15, a reflective test bar (geometric thickness ~3 μm) is embedded within the three samples to determine image resolution and contrast achievable when a polarizing annular objective was employed. Prepared samples are mounted onto a three-dimensional translation stage and the focus of the imaging objective O_2 is placed on the surface of the test bar. The degree of polarization, γ, for sample8 is measured to be 99.9 % for a circular imaging objective ($\varepsilon = 0$), which shows that depolarization caused by the imaging objective and other optical components in this experimental arrangement can be ignored.

Table 5.2 Parameters for the samples used in the reflection scanning optical microscope

Sample number	Scattering medium	Cell thickness, d (μm)	SMFP length, l_s (μm)	Anisotropy value, g	Optical thickness, n ($2d/l_s$)
8	Distilled water	120	–	–	–
9	0.48 μm polysty. microsph. in water	120	38.4	0.81	6.25
10	0.48 μm polysty. microsph. in water	120	19.2	0.81	12.5

All parameters are determined by Mie theory [12]

5.3.2 Effect of an Annular Imaging Objective

The suppression of diffusing photons by a polarizing annular objective can be seen from the measured dependence of the degree of polarization, γ, on the radius the central obstruction, ε, for sample9 and sample10 (Fig. 5.16a), while Fig. 5.16b illustrates the polarization intensity ratio, I_p/I_s, for the collected signal. The measured degree of polarization, γ, for the two samples is 0.52 (or $I_p/I_s = 3.2$) and 0.33 (or $I_p/I_s = 2.0$), respectively, when a circular objective ($\varepsilon = 0$) is used for imaging. The lower degree of polarization, γ, and polarization intensity ratio, I_p/I_s, for sample10 results from its larger optical thickness, n (see Table 5.2). Therefore, each photon propagating through sample10 has experienced on average twice as many scattering events as sample9 and therefore stronger depolarization than photons propagating in sample9.

To help demonstrate the performance of the central obstruction size, ε, on the suppression of scattered photons, the percentage improvement parameters, P_γ and

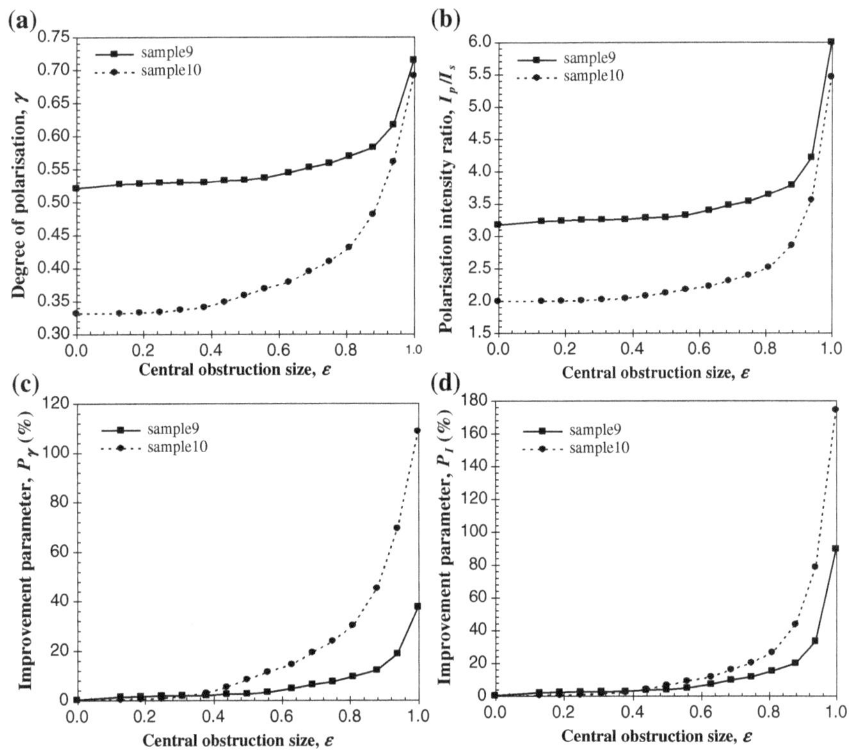

Fig. 5.16 Dependence of the degree of polarization, γ (**a**), the polarization intensity ratio, I_p/I_s (**b**), and the percentage improvement parameters, P_γ (**c**) and P_I (**d**) on the radius of the central obstruction of an annular imaging objective, ε. The numerical aperture for the imaging objective O_2 is 0.75. Reprinted with permission from [12], 1998, Cambridge University Press

P_I, defined by (5.2) and (5.3) are used. The parameters P_γ and P_I are illustrated in Fig. 5.16c, d for the two samples, respectively.

An appreciable increase in the degree of polarization, γ, is seen in Fig. 5.16a when a thin polarizing annular objective is employed, i.e., when the central obstruction size, ε, approaches unity. The degree of polarization, γ, for an annular objective of $\varepsilon \approx 0.94$ was 0.62 ($I_p/I_s = 4.2$) and 0.56 ($I_p/I_s = 3.5$) for the two samples, respectively. When compared with the circular objective ($\varepsilon = 0$), the increase in the degree of polarization, γ, is 0.1 (or $P_\gamma \approx 19$ %) and 0.23 (or $P_\gamma \approx 70$ %) for the two samples, respectively, while the corresponding improvement in the polarization intensity ratio, I_p/I_s, is 1.0 (or $P_I \approx 31$ %) and 1.5 (or $P_I \approx 75$ %) for the two samples. This result clearly demonstrates that if a thin polarizing annular objective is used for imaging, the weighting of scattered photons is efficiently reduced (i.e., I_p/I_s increases).

In Fig. 5.16 the degree of polarization, γ, monotonically increases with the size of the central obstruction, ε, unlike the situation in the transmission optical microscope in which case there is a weak dip on the γ–ε curve (see Fig. 5.7a). This difference is caused by the larger anisotropy value, g; a large anisotropy value, g, corresponds to the larger proportion of forward diffuse photons. As a result, a significant proportion of the forward diffuse photons are scattered out of the collection cone of the imaging objective upon reflection from the test bar surface. Thus the effect of the polarization annular objectives on the suppression of scattered photons becomes dominant earlier.

5.3.3 *Effect of the Numerical Aperture of an Objective*

Figure 5.17a shows the effect of the numerical aperture on the degree of polarization, γ, for sample9 and sample10 in the reflection arrangement, while Fig. 5.17b illustrates the polarization intensity ratio, I_p/I_s, for the collected signals. The two percentage improvement parameters, P_γ and P_I, defined in (5.4) and (5.5), are used to help demonstrate the performance of the numerical aperture on the suppression of scattered photons. The parameters P_γ and P_I are illustrated in Fig. 5.17c, d for the two samples, respectively.

It is clearly seen from Fig. 5.17 that the degree of polarization, γ, increases appreciably for both samples when the numerical aperture (NA) is decreased. The difference in the degree of polarization, γ, between NA = 0.75 and NA = 0.11 is 0.11 (or $P_\gamma \approx 21$ %) and 0.05 (or $P_\gamma \approx 13$ %) for sample9 and sample10, respectively. For the same conditions, the improvement in the polarization intensity ratio, I_p/I_s is 1.24 (or $P_I \approx 39$ %) and 0.22 (or $P_I \approx 11$ %) for sample9 and sample10, respectively. This result confirms that a trade-off between resolution and suppression ability should also be considered when a reflection optical arrangement is used for imaging through turbid media.

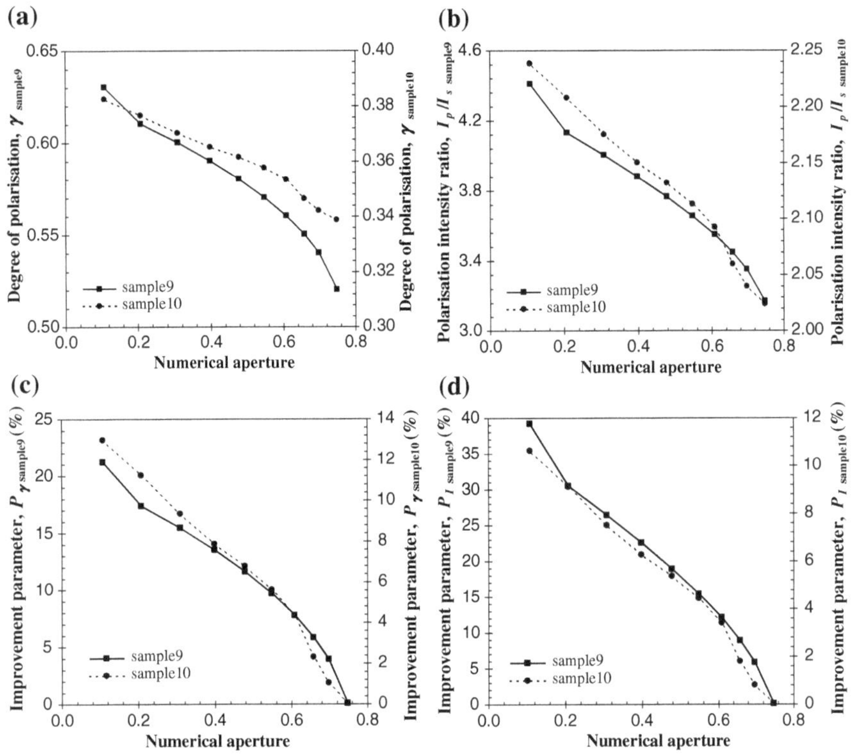

Fig. 5.17 Dependence of the degree of polarization, γ (**a**), the polarization intensity ratio, I_p/I_s (**b**), and the percentage improvement parameters, P_γ (**c**) and P_I (**d**) on the numerical aperture of the imaging objective O_2

5.3.4 Imaging with Circular and Annular Objectives

Like in the case for the transmission microscope, an optimal annular objective with a central obstruction size, $\varepsilon \approx 0.94$ is chosen to perform all imaging since this annular objective has a good suppression ability and high signal-to-noise ratio, and the incident power from the laser is attenuated with the use of a variable neutral density filter (Fig. 5.15) for the circular imaging objective ($\varepsilon = 0$) such that the illumination power at the focus is equal to the power at the focus for an annular objective ($\varepsilon \approx 0.94$). The power at the focus is approximately 110 μW. With a fixed incident power at the focus, a direct comparison of signal-to-noise ratio can be obtained for the two cases.

The test bar embedded within sample8 is imaged with a polarizing circular objective ($\varepsilon = 0$) to demonstrate the best image resolution and contrast achievable in our experiment (Fig. 5.18). The image in this case is constructed purely by unscattered photons obeying the diffraction theory of light [10, 11]. The sharpness of

Fig. 5.18 Image of a test bar embedded within sample8 (filled with distilled water), when a circular imaging objective (ε = 0) of numerical aperture 0.75 is employed. The intensity of the image is normalized to unity. Reprinted with permission from [12], 1998, Cambridge University Press

the edge, α, defined by the distance between 10 and 90 % intensity points of a fitted edge response, can be used as a measure of transverse image resolution. For example, at the point 'A' in Fig. 5.18, the transverse resolution, α, is approximately 3.1 μm. It is seen from Fig. 5.18 that the test bar has different line widths and variable reflectivity indicated by the levels of brightness. These properties can also be used as resolvability markers when the test bar is imaged through sample9 and sample10.

Figure 5.19 shows images of the test bar embedded in sample9 (Fig. 5.19a, b) and sample10 (Fig. 5.19c, d). In Fig. 5.19a, c a polarizing circular objective is used, while in Fig. 5.19b, d a polarizing annular objective ($\varepsilon \approx 0.94$) is used. The image of the test bar in Fig. 5.19a is smeared and the fine structure of the bar is washed out due to the strong contribution of scattered light. A significant improvement in resolution is seen with the introduction of a polarizing annular objectives (Fig. 5.19b), which is indicated by the reduced smearing and finer details (see the small square with a hole). The transverse resolution, α, in Fig. 5.19b at point 'A' is improved from 23.9 to 12.2 μm when compared with that in Fig. 5.18a, which corresponds to an improvement in transverse resolution, α, of approximately 49 %. It is noticeably difficult to see the test bar when the optical thickness, n, is increased (sample10) due to the poor contrast (Fig. 5.19c). This phenomenon can be attributed to the large number of scattered photons collected via the circular objective (ε = 0). Similar to the situation in Fig. 5.19b, the introduction of an annular objective ($\varepsilon \approx 0.94$) provides an image with sharper contrast and reduced blurring (Fig. 5.19d). In Fig. 5.19d, the transverse resolution, α, at point 'A' is improved from 46.8 to 35.1 μm when compared with that in Fig. 5.19c, which corresponds to an improvement in transverse resolution of approximately 25 %. These results quantitatively confirm that a thin imaging annular objective confines unscattered photons to a small region on the objective plane, which has the effect of reducing the weighting of scattered photons that contribute to the image.

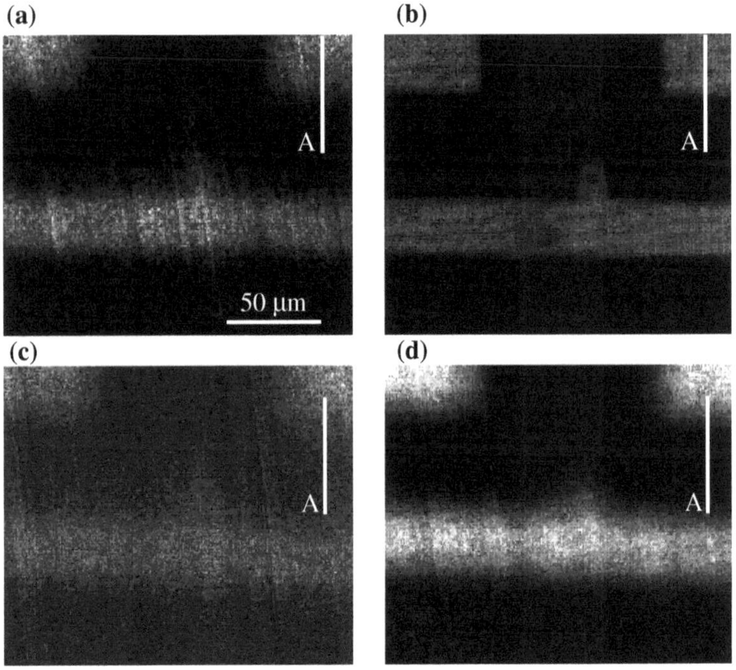

Fig. 5.19 Images of a test bar embedded in sample9 (**a**) and (**b**) and sample10 (**c**) and (**d**) when a circular imaging objective ($\varepsilon = 0$) with parallel polarization (**a**) and (**c**) and an annular imaging objective ($\varepsilon \approx 0.94$) with parallel polarization (**b**) and (**d**) are employed. The numerical aperture of the imaging objective in both cases is 0.75. The intensity of the images is normalized to unity. Reprinted with permission from [12], 1998, Cambridge University Press

5.4 Resolution in an Optical Microscope

According to the discussion in Sects. 5.2 and 5.3, it can be seen that using an objective in a microscopic imaging system raises the question of what numerical aperture of an objective is suitable for imaging. According to the imaging theory based on Born's approximation [10, 11], which ignores the multiple scattering in a turbid medium, a high numerical aperture objective lens can provide high diffraction-limited resolution. Born's approximation is applicable to the case that the optical thickness, n, defined as the thickness of a turbid medium divided by the scattering mean free path-length, is less than 1. On the other hand, a low numerical aperture objective can suppress scattered photons which statistically travel at high angles. Both arguments are based on the assumption that ballistic light is dominant in forming an image. When a turbid medium is thick, e.g., when $n > 10$, unscattered light may be too weak to be detected, particularly, in the presence of detector noise. In this situation, scattered light has to be included in creating an image. An important question raised here is what role the scattered light plays in forming an image. To answer this question, the relationship of scattered photons to image resolution should be investigated. In this section, we investigate the role of scattered light in constructing an image under a microscope, and propose a criterion for determining the efficiency of a gating method.

5.4.1 Resolution Contributed by Scattering Photons

The Monte Carlo simulation method, described in Chap. 3, is adopted to study light propagation through a turbid medium. The advantage of using the Monte Carlo simulation method is that it can track the change in physical properties of a photon during each scattering event. Therefore, it can predict the behavior of the scattered light and its role in constructing an image. In order to identify how many scattering events a photon has experienced before it leaves a turbid medium, the number of scattering events, N, is recorded during the simulation process.

A schematic diagram of a transmission scanning optical microscope is shown in Fig. 3.2. v_d is the normalized pinhole size [13]. Let us consider a thin sharp edge embedded in a scattering slab of thickness (d) 180 μm, which consists of 0.48 μm beads suspended in water. According to Mie scattering theory, the anisotropy value g of the scattering medium is 0.81 for a wavelength of 0.633 μm (see Table 2.1). From the image intensity of the sharp edge scanned in the x direction, the transverse resolution, Γ, is defined as the distance between the 90 and 10 % intensity points. All photons collected by the detector are divided into N groups according to the number of scattering events that a photon experiences. In this way, the transverse resolution, Γ_N, of an image constructed by each individual group of photons can be determined.

The transverse resolution, Γ_N, shown in Fig. 5.20a, increases with the number of scattering events, N, for a given value of the optical thickness. It is seen that once a photon is scattered, it contributes to an image with resolution much lower than the diffraction-limited resolution, which is less than 1 μm for a high numerical aperture objective. For example, photons that are scattered only once construct an image with resolution larger than 14 μm. This feature shows that an image with resolution near the diffraction limit can only be obtained if the unscattered light is dominant in the signal collected by the detector. In other words, in order to obtain an image with diffraction-limited resolution, most of the scattered photons have to be suppressed via a gating process.

However, the strength of the unscattered light decreases exponentially as the optical thickness, n, increases. In this situation, the unscattered signal is too weak to be detected compared with the noise generated in an imaging system. So when an object is embedded in a thick turbid medium, the scattered light has to be considered as part of the signal in constructing an image. This feature can be seen from the photon number distribution, ρ_N, normalized by the total number of input photons, as a function of the number of scattering events, N, for different values of the optical thickness (Fig. 5.20b). As can be observed, when a turbid medium becomes

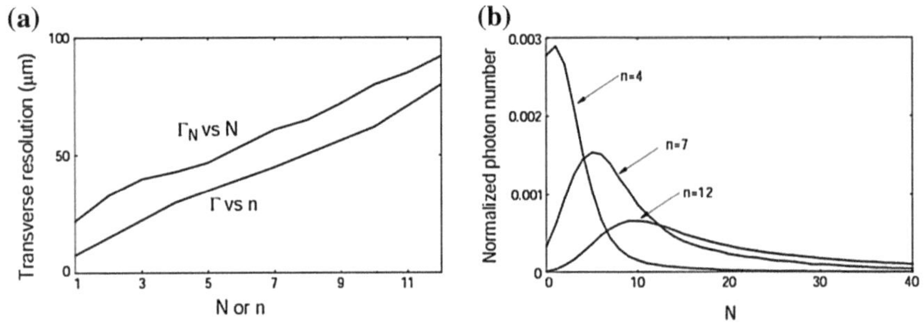

Fig. 5.20 **a** Image resolution as a function of the number of scattering events, N (n = 12) or the optical thickness, n. **b** Normalized photon number distribution, ρ_N, as a function of the number of scattering events, N, for different values of the optical thickness, n. NA = 0.25, $v_d \rightarrow \infty$. Reprinted with permission from [4], 1998, Optical Society of America

thick, the number of unscattered photons drops quickly, while the number of scattered photons increases, particularly those that are scattered multiple times. Due to the increase of the multiply scattered photons in the total number of photons, image resolution degrades quickly as a turbid medium becomes thick. The image resolution, Γ, as a function of the optical thickness, n, of a turbid medium is shown in Fig. 5.20a. As expected, the image resolution, Γ, is much lower than the diffraction-limited resolution even for n = 1 and becomes worsened as a turbid medium becomes thicker.

5.4.2 Effects of the Numerical Aperture and the Pinhole Size

In an optical microscope, for example, in a confocal microscope, the aperture size of objectives along with the size of the detector pinhole act as angular gates for suppressing scattered photons. Before a pinhole mask starts to play its role, photons emerging from a turbid medium first go through a preselection process determined by the aperture of objectives. In Fig. 5.21a, image resolution, Γ, as a function of the optical thickness, n, for different values of the numerical aperture of objectives is illustrated. It is noticed that an improvement in image resolution occurs when a low numerical aperture objective is used. A low numerical aperture objective rejects more photons traveling at high angles than a high numerical aperture objective. These photons statistically experience more scattering events and therefore contribute to an image with lower resolution. As a result, image resolution is higher for a lower numerical aperture objective. It should be mentioned that a high numerical aperture objective gives a narrow diffraction spot at its focus, and therefore gives a high-resolution image if scattered photons are negligible. But when scattering is strong, a high numerical aperture objective collects more scattered photons emerging from a turbid medium. Therefore, for a high numerical aperture objective,

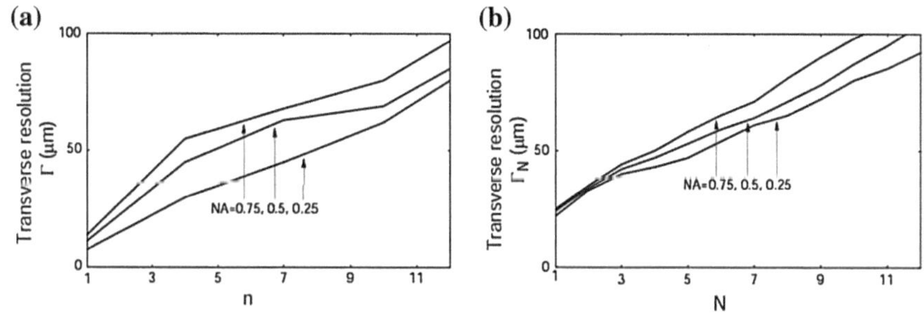

Fig. 5.21 **a** Image resolution, Γ, as a function of the optical thickness n for different values of the numerical aperture of objectives ($v_d \rightarrow \infty$). Image resolution, Γ_N, for different values of the numerical aperture of the objective (n = 12, $v_d \rightarrow \infty$). Reprinted with permission from [4], 1998, Optical Society of America

the degradation of image resolution caused by scattered photons is severe if no other gating mechanism is used.

The improvement in resolution, caused by the aperture of an objective, comes from two aspects: one is from the reduction of scattered photons experiencing a given number of scattering events at high angles, and the other is from the change in the photon number distribution, ρ_N. To explain these two aspects further, we show, in Fig. 5.21b, image resolution, $\Gamma_{N,}$ for different values of the numerical aperture of objectives. These curves clearly show that for a given number of scattering events, photons traveling at high angles carry less information of the object and can be suppressed by the aperture of an objective.

The normalized photon number distribution, ρ_N, for different values of the numerical aperture of an objective is shown in Fig. 5.22a. It is seen that a high numerical aperture objective collects more scattered photons than a low numerical aperture objective. But the profiles of the normalized photon number distribution, ρ_N, are similar in all cases. For example, the ratio ρ_{12}/ρ_{50} is 4.84, 4.7, and 4.55 for numerical aperture (NA) = 0.25, 0.5, and 0.75, respectively. It is clear from Figs. 5.21b and 5.22a that using a low numerical aperture does show a sound improvement in the first aspect but the change in the second aspect is not so significant.

A pinhole mask in a confocal microscope has a strong effect on suppressing scattered photons. How a pinhole selects photons depends on the deviation of a scattered photon from the path of unscattered photons. The image resolution, Γ, as a function of the optical thickness, n, for different pinhole sizes is shown in Fig. 5.23a. It is noticed that when a finite-sized pinhole is placed in front of the detector, image resolution is improved significantly compared with the case without a pinhole. A pinhole can suppress scattered photons because they are statistically distributed away from the center of the detector plane and the distance from the center reflects how much a scattered photon deviates from the path of ballistic photons. Therefore, a finite-sized pinhole can reduce the degradation of image resolution caused by scattered photons.

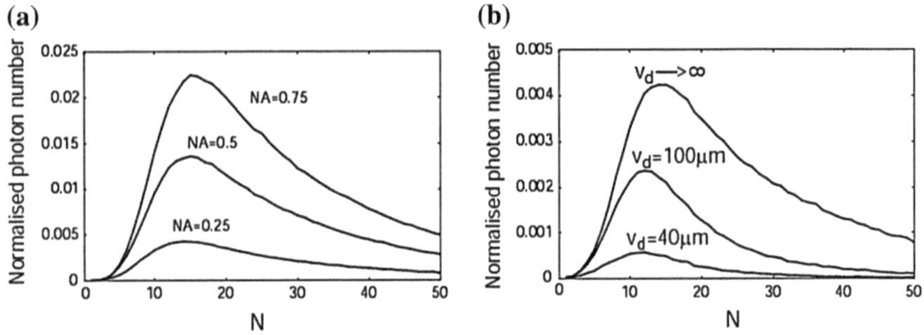

Fig. 5.22 **a** Normalized photon number distribution, ρ_N, as a function of the number of scattering events, N, for different values of the numerical aperture of objectives ($n = 12$, $v_d \rightarrow \infty$). **b** Normalized photon number distribution, ρ_N, as a function of the number of scattering events, N, for different pinhole sizes ($n = 12$, NA = 0.25). Reprinted with permission from [4], 1998, Optical Society of America

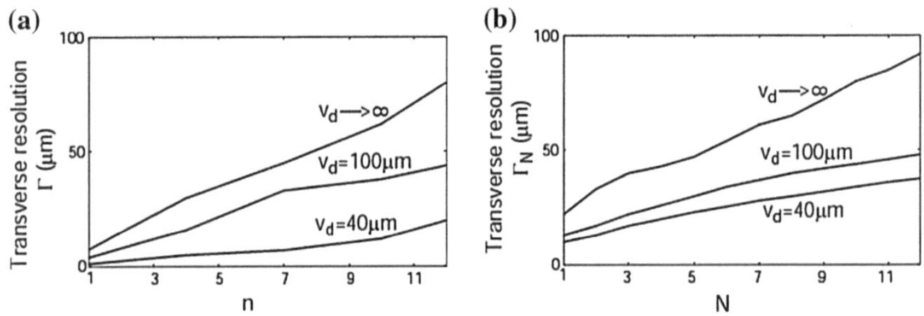

Fig. 5.23 **a** Image resolution, Γ, as a function of the optical thickness n for different pinhole sizes (NA = 0.25). **b** Image resolution, Γ_N, for different pinhole sizes ($n = 12$, NA = 0.25). Reprinted with permission from [4]

To further analyze the effect of the pinhole size, we show the transverse resolution, Γ_N, in Fig. 5.23b. It is seen that when a finite-sized pinhole is used, the resolution of an image formed by photons scattered N times is improved noticeably. Even when $v_d = 100$ μm, the improvement in resolution is significant compared with the case without a pinhole. This finding shows that a pinhole selectively reduces the number of scattered photons for a given number of scattering events, N. In other words, a pinhole actually suppresses scattered photons which are distributed further away from the center of the detector plane and carry less information of the object. The photon number distribution, ρ_N, emerging from the turbid medium and collected by a pinhole of different sizes is shown in Fig. 5.23b. As expected, scattered photons are reduced significantly if a small pinhole is used. It can also be noted that the reduction of scattered photons is more significant when the number of scattering events, N, is larger. This feature is reflected by the fact that the ratio ρ_{12}/ρ_{50} is 4.84, 20.5, and 28 for pinholes of $v_d \rightarrow \infty$, $v_d = 100$ μm, and

v_d = 40 μm, respectively. Thus the use of a finite-sized pinhole reduces the weighting of more scattered photons. Therefore, the overall image resolution Γ can be improved since photons experiencing more scattering events contribute to an image with lower resolution. It is clear from Figs. 5.23b and 5.22b that a finite-sized pinhole not only reduces the number of scattered photons according to their path deviation in the detector plane but also changes the weighting of the scattered photons in the total number of photons collected by a detector.

5.4.3 Relationship of Resolution to Signal Level

According to the discussion in Sects. 5.4.1 and 5.4.2, an image of high resolution can be acquired by the use of a low numerical aperture objective or a pinhole of small size, which leads to low signal strength. This trade-off between image resolution and signal strength actually occurs in most gating methods. A criterion for determining the efficiency of a gating method is proposed in Fig. 5.24.

In Fig. 5.24a, the image resolution, Γ, as a function of signal strength, defined as the ratio of the number of photons collected by a detector to the number of input photons, is illustrated for different values of the numerical aperture of an objective. It is seen that for given resolution, the signal strength for a high numerical aperture objective is higher than that for a low numerical aperture objective. More interestingly, for a given signal strength, image resolution, Γ, is much higher for a high numerical aperture objective. To understand this feature, we consider two points A and B in Fig. 5.24a, which correspond to two combinations of the numerical aperture and the pinhole size. The signal strength in both cases is the same. For point A, an objective of NA = 0.25 and a pinhole of v_d = 4,000 μm are used, while for point B, an objective of NA = 0.75 and a pinhole of v_d = 100 μm are used. The image resolution, Γ, in these two cases is 59 and 42 μm, respectively, which shows a better resolution in the latter case. The reason for this phenomenon can be drawn from the photon number

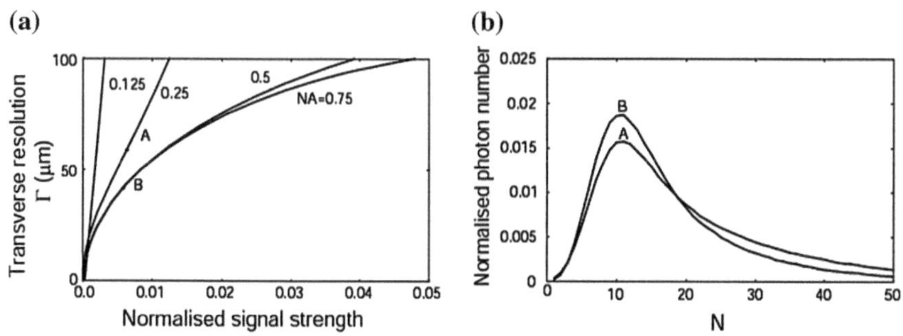

Fig. 5.24 **a** Image resolution, Γ, as a function of signal strength for different values of the numerical aperture of objectives (n = 12). **b** A comparison of the normalized photon number distribution, ρ_N between cases *A* and *B*. Reprinted with permission from [4], 1998, Optical Society of America

distribution, ρ_N, in the two cases, which is illustrated in Fig. 5.24b. It is noted from Fig. 5.24b that although the total number of photons collected by the detector is almost the same in the two cases, the weighting of the less scattered photons in case B is larger than that in case A. This feature shows that a small pinhole in case B is more efficient in suppressing highly scattered photons than a low numerical aperture objective in case A. The results shown in Fig. 5.24 suggest that the gating mechanism provided by an objective, which removes scattered photons based on their propagation angle, is not as efficient as that by a pinhole, which selects photons based on the deviation from the path of ballistic photons.

It is worthwhile to compare, at least qualitatively, the new polarized annular objective gating method with other gating methods currently used in microscopic imaging through turbid media. A confocal pinhole that plays an angle-gating role has been proved to be a useful mask to suppress scattered photons. However, the signal level in confocal microscopy may be reduced dramatically in imaging through turbid media [14]. In addition, a small pinhole may lead to pronounced coherent noise due to the purely coherent imaging property in confocal microscopy [14]. These two problems do not exist in the system discussed in the case of annual objectives. In fact, polarized annular objectives combined with a large pinhole could result in a better imaging performance by suppressing scattered photons. In this respect, incorporating polarized annular objectives into optical coherence tomography would be another potential method for efficient suppression of scattered photons. Since the angle-gating method in this chapter does not involve optical fibers and interference arrangements, the new method may be more easily implemented in a practical microscope than optical coherence microscopy [15].

References

1. M. Gu, X. Gan, Monte-Carlo simulation for confocal imaging through highly scattering media. Scanning **19**, 148 (1997)
2. X. Gan, M. Gu, Modified Monte Carlo simulation of multi-dimensional photon distributions for microscopic imaging. Optik **108**, 129 (1998)
3. X. Gan, S. Schilders, M. Gu, Combination of annular aperture and polarisation-gating methods for efficient microscopic imaging through a turbid medium: theoretical analysis. Microsc. Microanal. **3**, 495 (1997)
4. X. Gan, S. Schilders, M. Gu, Image formation in a turbid medium under a microscope. J. Opt. Soc. Am. A **15**, 2052 (1998)
5. S.P. Schilders, X.S. Gan, M. Gu, Microscopic imaging through a turbid medium using annular objectives for angle-gating. Appl. Opt. **37**, 5320 (1998)
6. S.P. Schilders, X.S. Gan, M. Gu, Efficient suppression of diffusing photons using polarising annular objectives for microscopic imaging through turbid media. Bioimaging **6**, 92 (1998)
7. R.G. Jensen, *Handbook of Milk Composition* (Academic Press, San Diego, 1995)
8. S.P. Morgan, M.P. Khong, M.G. Somekh, Effects of polarization state and scatterer concentration on optical imaging through scattering media. Appl. Opt. **36**, 1560 (1997)
9. H. Wabnitz, H. Rinneberg, Imaging in turbid media by photon density waves: spatial resolution and scaling relations. Appl. Opt. **36**, 64 (1997)
10. M. Gu, *Advanced Optical Imaging Theory* (Springer, Heidelberg, 2000)

11. M. Born, E. Wolf, *Principles of Optics* (Pergamon, New York, 1980)
12. S.P. Schilders, X. Gan, M. Gu, Image enhancement in a reflection optical microscope by suppression of diffusing photons using polarising annular objectives. Microsc. Microanal. **4**, 415 (1998)
13. M. Gu, *Principles of Three-dimensional Imaging in Confocal Microscopes* (World Scientific, Singapore, 1996)
14. M. Gu, T. Tannous, C.J.R. Sheppard, Effect of an annular pupil on confocal imaging through highly scattering media. Opt. Lett. **21**, 312 (1996)
15. M. Hee, J. Izatt, E. Swanson, J. Fujimoto, Femotsecond transillumination tomography in thick tissues. Opt. Lett. **18**, 1107 (1993)

Chapter 6
Polarization-Gating Mechanism

Abstract The angle-gating mechanism discussed is Chap. 5 is one possible method of suppressing scattered photons in imaging through turbid media. However, the limitation of the angle-gating mechanism is that it requires a significant proportion of unscattered photons to be available for detection. Thus, as the scattering event increases the signal-to-noise ratio of the detected images is substantially reduced. To solve this problem, a polarization-gating method can be adopted. This chapter is dedicated to the experimental and theoretical understanding of the polarization-gating mechanisms in an optical microscope (Schilders et al. Appl Opt 37:4300,1998; Schilders et al. Opt Commun 157:238, 1998; Gan et al. J Opt Soc Am A 16:2177, 1999). Section 6.1 discusses the principles of polarization-gating. The performance of the polarization-gating mechanisms on image quality for different sized scatterers is experimentally presented for a reflection optical microscope in Sect. 6.2. With the use of the Monte Carlo simulation model, the polarization-gating performance in reflection and transmission optical microscopes is investigated in Sects. 6.3 and 6.4, respectively. Finally, the effective point spread function together with image resolution and signal level in a transmission optical microscope is presented in Sect. 6.5.

6.1 Principle of Polarization-Gating

Unscattered or less scattered photons maintain the same or similar degree of incident polarization, and carry more information of an object embedded within turbid media than multiply scattered photons [1–3]. It is therefore possible to separate the unscattered or less scattered photons with a polarization detection mechanism. To this end, one uses a polarized beam for illumination. There are a number of methods for polarization-gating. The first method is based on an analyzer that is placed to be perpendicular to the direction of the polarizer. In this case, only unpolarized photons are selected, which form the perpendicular polarization-gating (PPG) mechanism. An image built upon scattered photons shows low resolution. The second mechanism is the method when the analyzer is parallel with the direction of the

M. Gu et al., *Microscopic Imaging Through Turbid Media*,
Biological and Medical Physics, Biomedical Engineering,
DOI 10.1007/978-3-662-46397-0_6

polarizer, in which case, unscattered and/or less scattered photons are used for imaging, which is called conventional polarization-gating (CPG). Image quality can be further improved if the differential polarization-gated intensity, which can be derived by subtracting the perpendicular polarization-gated intensity from the parallel polarization-gated intensity, is used to construct an image. This mechanism is based on the fact that depolarized scattered photons contribute to both the perpendicular polarization-gated intensity and the parallel polarization-gated intensity.

The speed of depolarization of scattered photons is related to the directional change of the incident light upon scattering and is determined by the anisotropy value, g. For small scattering particles, the directional change of scattered photons is large due to the small anisotropy value, g, which results in a significant change in the polarization state. Therefore, polarization-gating and in particular, differential polarization-gating (DPG) is more effective when turbid media with small scattering particles are imaged.

For a transillumination imaging system, Morgan et al. demonstrated that the performance of DPG depends on the characteristics of turbid media, such as the size of scatterers and the thickness of turbid media [4, 5]. The scatterer size determines the speed of depolarization, while the sample thickness determines the magnitude of the depolarization. It has been demonstrated that image resolution of 0.7 mm, for an edge object embedded within a semi-skimmed milk suspension (optical thickness $\approx$ 18), can be obtained with differential polarization-gating when a parallel beam probe of approximately 1 mm in diameter is used in a transillumination imaging system [5]. Other researchers have also demonstrated that an improvement in image quality is possible when DPG is used in optical coherence tomography [6] and in fluorescence microscopy [7].

In the transillumination optical arrangement, the image resolution is determined by the size of the parallel beam probe. Therefore, to improve image resolution further, it is necessary to employ a microscope objective. The polarization-gating mechanism is one of the methods which can be applied to microscopic imaging for efficiently selecting the unscattered or the less scattered photons.

6.2 Polarization-Gating in a Reflection Optical Microscope

In this section, a detailed investigation into the effect of polarization-gating (conventional and differential) mechanisms used in a reflection optical microscope for imaging through turbid media is presented. The effect of the scattering particle size in turbid media on microscopic imaging is experimentally studied in order to determine the resolution achievable when polarization-gating mechanisms are employed. Image resolution for a 48 μm wide bar embedded within turbid media is measured as a function of the optical thickness for four different types of polystyrene microspheres suspended in water. To demonstrate the effectiveness of polarization-gating in more realistic tissue-like turbid media, image resolution of a 48 μm wide bar embedded within semi skimmed-milk suspensions is measured.

6.2.1 Experimental Details

A reflection scanning optical microscope employing polarization-gating mechanisms is depicted in Fig. 6.1 [2]. A 7 mW Melles Griot (model No. 05-LHP-153) He–Ne laser (λ = 632.8 nm) was used as the light source. The beam from the laser source is expanded and collimated by an objective O_1 (numerical aperture NA = 0.25) and a lens L_1 (focal length, f = 200 mm; diameter, D = 25 mm) to match the entrance aperture of an imaging objective O_2 (Zeiss FLUAR 20×/0.75). Scattered and unscattered photons collected via the imaging objective O_2 are separated into parallel and perpendicular polarization components by analysers LP_2 and LP_3 relative to the incident polarization state defined by the linear polariser LP_1. The parallel and perpendicular polarization components are then focused via lenses L_2 and L_3 (D = 25 mm) onto two identical large area (1 cm^2) detectors DT_1 and DT_2 which record time-averaged signals, I_p (termed the conventional polarization-gated intensity) and I_s (termed the perpendicular polarization-gated intensity), respectively. The no-gated intensity, $I_p + I_s$, and the differential polarization-gated, intensity, $I_p - I_s$, can be calculated with the aid of a computer. The degree of polarization, γ, of the light reflected from a turbid sample can be calculated with the aid of (3.10).

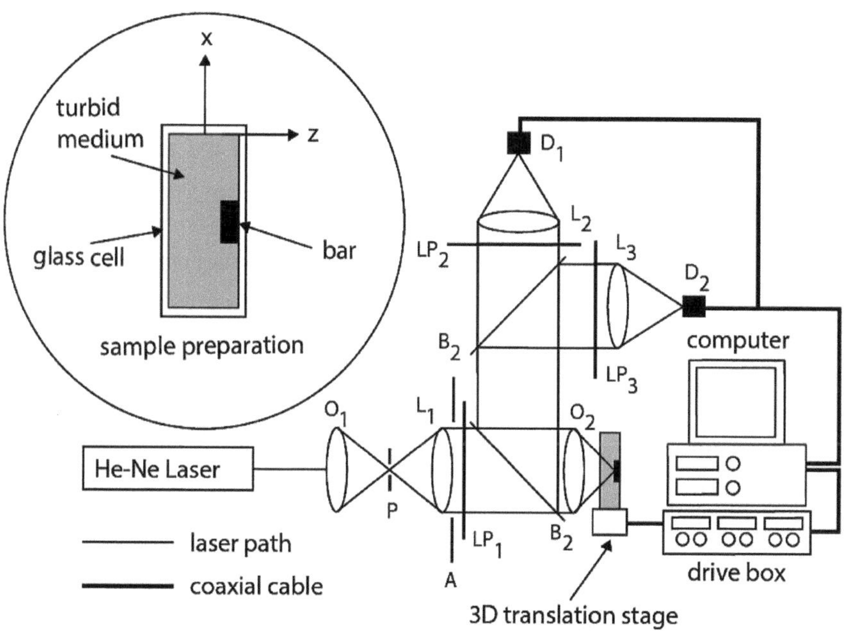

Fig. 6.1 Experimental reflection scanning optical microscope with differential polarization-gating mechanism. *A* aperture, *B* beamsplitter, *DT* detector, *L* lens, *LP* polariser, *O* objective, *P* pinhole. Reprinted with permission from [1], 1998, Optical Society of America

Four types of polystyrene microspheres suspended in water were respectively used as scattering particles in a turbid medium. The detailed scattering parameters for the four types of microspheres are given in Table 2.1. A number of samples are prepared for each type of polystyrene microspheres. The samples produced with each polystyrene microsphere size are grouped into sample sets A–D and their details are listed in Table 6.1. The range of the microsphere sizes chosen is consistent with that of the scattering particles in real biological samples which do not include large scatterers (>1 μm) [6, 8].

Since real tissue-like samples consist of scatterers of different sizes, it is important to demonstrate the effect of the polarization-gating mechanisms on image quality in such a turbid medium. To demonstrate image quality in a more realistic turbid medium a sample set of semi-skimmed milk is prepared (i.e. sample set E in Table 6.1). Semi-skimmed milk (fat content 3.6 %) is chosen since it does not include large scatterers (>1 μm), and has been used in transillumination imaging to simulate tissues which do not include large scatterers (>1 μm) [6, 8]. If the tissue includes large scatterers, polarization-gating may not be efficient as large scatterers (high anisotropy value, g) result in high forward scattering without appreciably changing the degree of polarization, γ, of the incident beam. The size of scattering particles in the milk suspension is approximately in a range of 0.1–1 μm, which is similar to that used in the case of polystyrene microspheres (samples sets A–D). According to Beer's law, the scattering mean free path length, l_s, of the semi skimmed milk was measured to be approximately 30 μm (see Fig. 5.1b). The anisotropy value, g, for the semi-skimmed milk suspension is assumed from the literature [9].

Each type of polystyrene microsphere is placed in a glass cell with lateral dimensions of 2 cm × 1 cm. The thickness of the glass cell, d, can be varied between 25 and 600 μm, producing an optical thickness of a turbid sample, $n = 2d/l_s$, for the reflection microscope. It should be noted that the maximum optical

Table 6.1 Parameters for the sample sets used in the reflection scanning optical microscope

Sample set	Scattering medium	Cell thickness, d (μm)	SMFP length, l_s (μm)	Anisotropy value, g	Optical thickness, n ($2d/l_s$)
A	0.107 μm polysty. microsph. in water	100–600	416.4[a]	0.076[a]	0.5–2.8
B	0.202 μm polysty. microsph. in water	100–600	70.2[a]	0.31[a]	2.8–17.1
C	0.48 μm polysty. microsph. in water	25–300	19.2[a]	0.81[a]	2.6–31.3
D	1.053 μm polytsy. microsph. in water	25–300	10.8[a]	0.92[a]	4.6–55.6
E	Semi-skimmed milk	50–300	30[b]	<0.4[c]	3.3–20

[a] The parameters are determined by Mie theory (see Table 2.1)
[b] The parameters are determined experimentally (see Fig. 5.1b)
[c] The parameters are determined from the literature [9]

Table 6.2 Parameters for the specific samples used in the reflection scanning optical microscope

Sample number	Scattering medium	Cell thickness, d (μm)	SMPF length, l_s (μm)	Anisotropy value, g	Optical thickness, n $(2d/l_s)$
1	Distilled water	100	–	–	–
2	0.107 μm polysty. microsph. in water	600	416.4[a]	0.076[a]	2.8
3	0.202 μm polysty. microsph. in water	100	70.2[a]	0.31[a]	2.8
4	Semi-skimmed milk	100	30[b]	<0.4[c]	6.7
5	Semi-skimmed milk	300	30[b]	<0.4[c]	20
6	1.056 μm polysty. microsph. in water	100	10.8[a]	0.92[a]	18.5

[a] The parameters are determined by Mie theory (see Table 2.1)
[b] The parameters are determined experimentally (see Fig. 5.1b)
[c] The parameters are determined from the literature [9]

thickness, n, achievable for a given turbid medium set is determined by the working distance of the imaging objective O_2 used. To qualitatively demonstrate the image performance of polarization-gating, a selected set of six samples are imaged in the experiments. The detailed parameters of the six samples (numbered 1–6) are illustrated in Table 6.2.

It should be pointed out that the value of the optical thickness, n, can be altered by changing either the geometric cell thickness, d, of a sample or the concentration of scatterers or both. For a given scatterer size, the time-averaged angular distribution of scattered photons is dependent only on the value of the optical thickness, n, although their spatial distribution is determined by both the optical thickness, n, and the scattering mean free path length, l_s [10]. The spatial distribution of scattered photons together with a finite-sized detector and the collection objective used in this experimental investigation introduces additional gating. As a result, increasing the geometric thickness, d, of a turbid medium affects the image resolution more appreciably than increasing the concentration of scatterers. In addition, increasing the sample thickness, d, is more realistic since it represents an increase in the penetration depth for a biological sample.

For the evaluation of image quality, a 48-μm wide metallic reflective bar (reflectivity ≈ 70 % at $\lambda = 632.8$ nm) is embedded within one of the turbid media described in Table 6.1. Prepared samples are mounted onto a three-dimensional translation stage and the sample cell is scanned along the x-direction with the focus of the imaging objective O_2 on the surface of the bar. Transverse resolution of the scanned bar image, α, is defined as the distance between the 10 and 90 % intensity points, measured from the responses after they are fitted by using a symmetric-double-sigmoidal (SDS) method. This definition of resolution is consistent with that used in the Monte Carlo simulation discussed in Sect. 5.4.

The transverse resolution of the bar image, α, and the degree of polarization, γ, for sample1 is 0.7 μm and 99.9 %, respectively, for an imaging objective, O_2, of

NA 0.75. The image in this case is constructed purely by unscattered photons obeying the diffraction theory of light and therefore demonstrates the maximum image resolution achievable in the experimental arrangement.

6.2.2 Effect of the Optical Thickness and Scatterer Size on Resolution

Figure 6.2a–c show typical images of a 48 μm bar embedded in sample2 with scatterers of diameter 0.107 μm (see Table 6.2) when CPG, PPG and DPG techniques are employed, respectively. It can be seen that the image constructed by DPG (Fig. 6.2c) has more detail when compared with the image constructed by CPG (Fig. 6.2a), which implies the effectiveness of DPG over CPG in suppressing scattered photons. It should also be noted that an image can be constructed with mainly scattered photons (Fig. 6.2b), although its resolution is lower than that of the images formed with CPG and DPG techniques. This feature is illustrated more clearly in Fig. 6.2d which shows typical edge responses (at the position marked by 'A') of the embedded bar and corresponding fitted SDS curves recorded with CPG (I_p), PPG (I_s) and DPG (I_{p-s}), for the images illustrated in Fig. 6.2a–c), respectively. The measured transverse resolution, α, for CPG (α_p), perpendicular polarization-gating (α_s) and differential polarization-gating (α_{p-s}) techniques was 6.7, 9.2 and 4.0 μm, respectively. These results demonstrate that once a photon is scattered, it can only contribute to an image with resolution lower than the diffraction limited resolution.

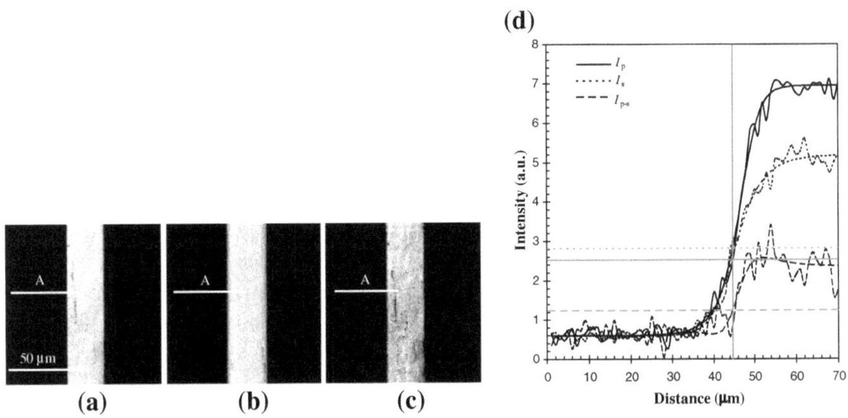

Fig. 6.2 Measured images for sample2 (Table 6.2) when CPG (**a**), PPG (**b**), and DPG (**c**) techniques are employed. The intensity of the images is normalized to unity. **d** Includes the measured edge responses corresponding to CPG (I_p), PPG (I_s), and DPG (I_{p-s}) techniques. The *horizontal gray lines* represent the intensity for I_p (*solid*), I_s (*dot*) and I_{p-s} (*dashed*), at the bar edge position indicated by the *vertical gray line*. The NA of the imaging objective is 0.25. Reprinted with permission from [2], 1998, Elsevier

The measured transverse resolution as a function of the optical thickness, n, for sample set A with scatterers of diameter 0.107 μm (see Table 6.1) using imaging objectives with NA = 0.25 and NA = 0.75 is illustrated in Fig. 6.3a, b, respectively, for no-gating (NG), (α_{p+s}), CPG (α_p), PPG (α_s), and DPG (α_{p-s}) techniques. It is evident from Fig. 6.3a, b that using CPG (α_p) is better than NG (α_{p+s}) and that DPG (α_{p-s}) is supcrior to CPG (α_p).

According to diffraction theory [11], the image intensity at a sharp edge is 1/3 and 1/2 of the intensity far from the edge for partially coherent and incoherent

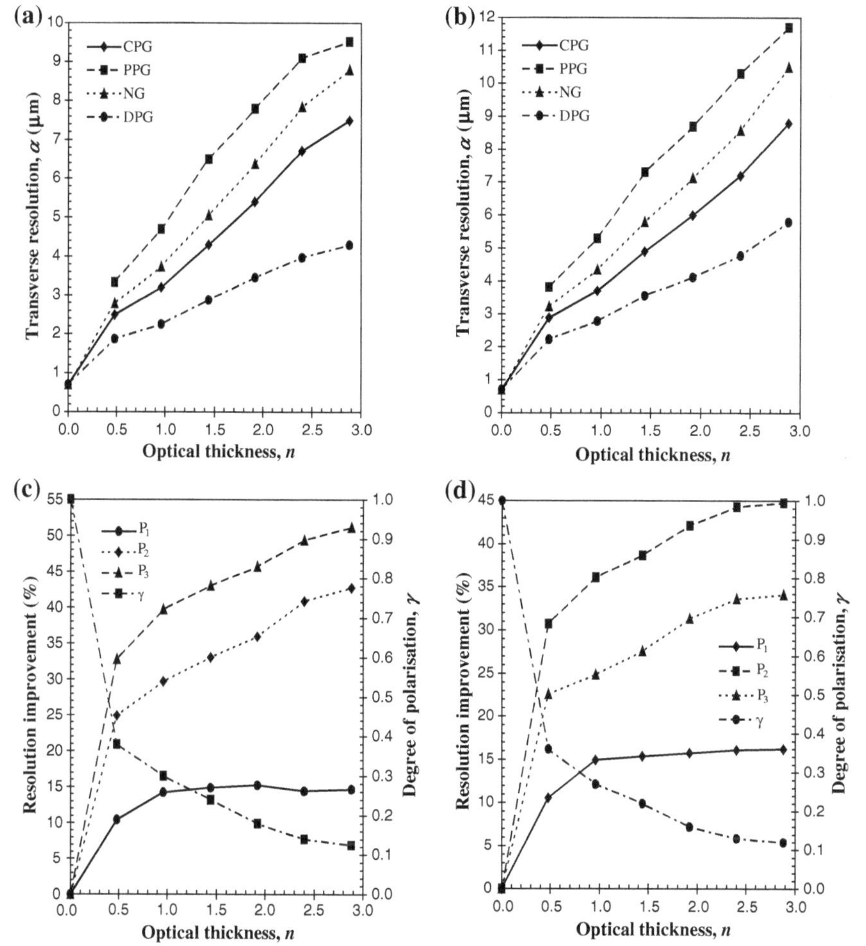

Fig. 6.3 Resolution performance for sample set A at NA = 0.25 (**a**) and (**c**) and NA = 0.75 (**b**) and (**d**). **a** and **b** Measured transverse resolution as a function of the optical thickness, n, for NG (α_{p+s}), CPG (α_p), PPG (α_s) and DPG (α_{p-s}). **c** and **d** Resolution improvement and the degree of polarization, γ, as a function of the optical thickness, n. A comparison is given between CPG and NG (P_1), DPG and NG (P_2), and DPG and CPG (P_3). Reprinted with permission from [2], 1998, Elsevier

imaging processes, respectively. In the case of CPG, the image is constructed mainly by unscattered photons and imaging is therefore partially coherent due to the use of a large area detector [11]. As a result, the image intensity at the edge is close to 1/3 of the intensity far from the edge (see the horizontal gray solid line in Fig. 6.2d). For PPG, the image is constructed by scattered photons (i.e., the incoherent component) and thus the intensity at the edge is close to 1/2 of the maximum intensity. In the DPG process, the unscattered component of the intensity is dominant due to the subtraction of the scattered photon component and therefore the intensity at the edge for DPG is close to 1/3 of the intensity far from the edge (see the horizontal gray dashed line in Fig. 6.2d). These three intensity points have been marked at the edge position represented by the vertical gray solid line in Fig. 6.2d.

To help illustrate the improvement in transverse resolution achieved with the implementation of a particular polarization-gating method, we introduce the following three improvement parameters. The improvement in transverse resolution between CPG (α_{p}) and NG ($\alpha_{\mathrm{p+s}}$) is defined as

$$P_1\big|_n = \frac{\alpha_{\mathrm{p+s}}\big|_n - \alpha_{\mathrm{p}}\big|_n}{\alpha_{\mathrm{p+s}}\big|_n}, \tag{6.1}$$

the improvement in transverse resolution between DPG ($\alpha_{\mathrm{p-s}}$) and NG ($\alpha_{\mathrm{p+s}}$) is defined as

$$P_2\big|_n = \frac{\alpha_{\mathrm{p+s}}\big|_n - \alpha_{\mathrm{p-s}}\big|_n}{\alpha_{\mathrm{p+s}}\big|_n}, \tag{6.2}$$

and the improvement in transverse resolution between CPG (α_{p}) and DPG ($\alpha_{\mathrm{p-s}}$) is defined as

$$P_3\big|_n = \frac{\alpha_{\mathrm{p}}\big|_n - \alpha_{\mathrm{p-s}}\big|_n}{\alpha_{\mathrm{p}}\big|_n}, \tag{6.3}$$

where n is the optical thickness.

The improvement in resolution is illustrated more clearly in Fig. 6.3c, d, which shows the percentage improvement in transverse resolution between CPG and NG (P_1), DPG and NG (P_2), and DPG and CPG (P_3). It is clearly seen from Fig. 6.3c, d that the improvement in resolution becomes large as the optical thickness, n, is increased. DPG has a strong effect when applied to this turbid medium because scattered photons are depolarized fast when the optical thickness, n, increases, as shown by the degree of polarization, γ (Fig. 6.3c, d), and also because there is still a significant proportion of unscattered light detectable (i.e., the optical thickness, n, is small). The reason for the higher improvement in resolution and a higher degree of polarization, γ, for, NA = 0.25, compared with those for NA = 0.75, is due to the stronger suppression of multiply scattered photons when the acceptance angle of an imaging objective is reduced (see Fig. 5.17). However, a higher NA imaging objective may be more effective for DPG, in terms of the percentage improvement in transverse resolution, when the optical thickness, n, is large and/or when the depolarization of each scattering event is weak. This effect will be discussed in more detail in Sect. 6.2.4.

Figure 6.5a illustrates the measured transverse resolution as a function of the optical thickness, n, for NG (α_{p+s}), CPG (α_p), PPG (α_s), and DPG (α_{p-s}) techniques when sample set B with scatterers of diameter 0.202 μm (see Table 6.1) and an imaging objective of NA 0.75 is used. It is evident from Fig. 6.5a, and has already been seen in Fig. 6.3a, b, that DPG (α_{p-s}) is superior to NG (α_{p+s}) and CPG (α_p). This improvement is illustrated more clearly in Fig. 6.5b, which shows the percentage improvement in transverse resolution between CPG and NG (P_1), DPG and NG (P_2), and DPG and CPG (P_3). A comparison of Figs. 6.3d and 6.5b shows that the improvement in transverse resolution between DPG and NG (P_2), in the latter case increases with the optical thickness, n (i.e., no steady-state value is reached). This phenomenon is because the depolarization of scattered photons in Fig. 6.5b is slower than that in Fig. 6.3d, when the optical thickness, n, increases, due to the larger anisotropy value, g, for the former case.

Images of an embedded bar in sample3 with scatterers of diameter 0.202 μm (the optical thickness, n, for this sample is equivalent to the images of sample2 (see Table 6.2 and Fig. 6.2) are shown in Fig. 6.4a–c for CPG, PPG, and DPG, respectively. A comparison of Figs. 6.2 and 6.4 demonstrates that the resolution for the latter case is significantly poorer, which is indicated by the lack of detail, by smearing and by noise. The degradation in resolution is first due to the increase in forward propagation of the scattered photons as a result of the larger anisotropy value, g, and secondly due to the collection of more scattered photons as a result of the larger NA (see Fig. 5.17).

Let us turn to sample set C with scatterers of diameter 0.48 μm. Under the same experimental condition as Figs. 6.3b, d and 6.5, transverse resolution as a function of the optical thickness, n, for NG (α_{p+s}), CPG (α_p), PPG (α_s), and DPG (α_{p-s}) techniques is shown in Fig. 6.6a. It can be seen from Fig. 6.6a that DPG (α_{p-s}) is superior to NG (α_{p+s}) and CPG (α_p) only when the optical thickness, n, is less than 30.

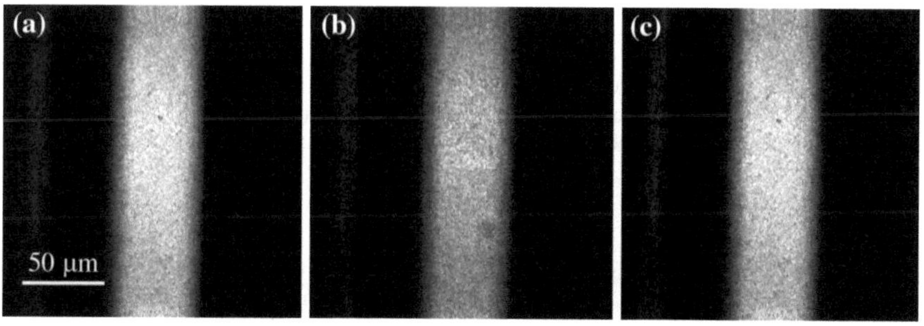

Fig. 6.4 Measured images for sample3 (Table 6.2) for CPG (**a**), PPG (**b**), and DPG (**c**) techniques. The intensity of the images is normalized to unity. The numerical aperture of the imaging objective is 0.75. Reprinted with permission from [2], 1998, Elsevier

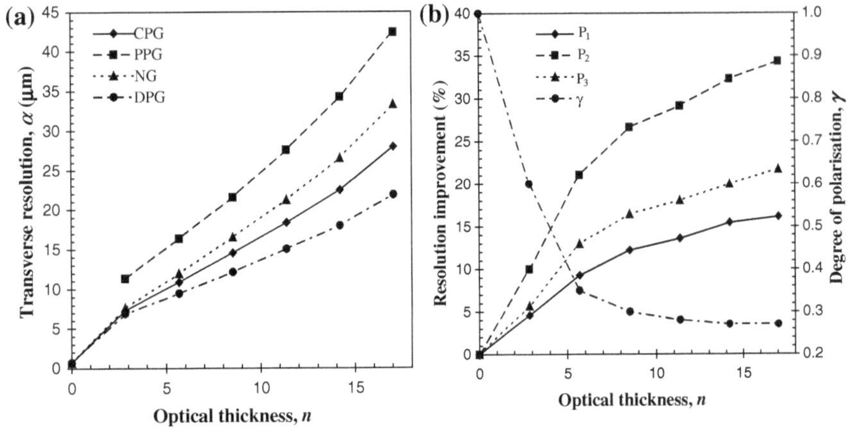

Fig. 6.5 Resolution performance for sample set B at NA = 0.75. **a** Measured transverse resolution as a function of the optical thickness, n, for NG (α_{p+s}), CPG (α_p), PPG (α_s), and DPG (α_{p-s}). **b** Resolution improvement and the degree of polarization, γ, as a function of the optical thickness, n. A comparison is given between CPG and NG (P_1), DPG and NG (P_2) and DPG and CPG (P_3). Reprinted with permission from [2], 1998, Elsevier

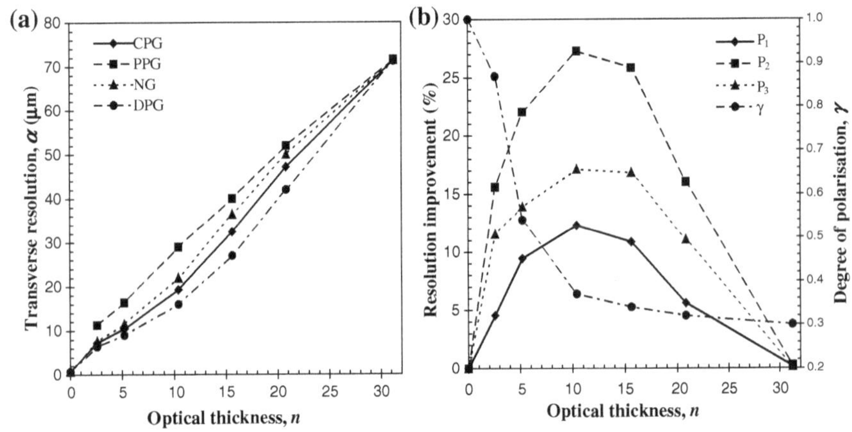

Fig. 6.6 Resolution performance for sample set C at NA = 0.75. **a** Measured transverse resolution as a function of the optical thickness, n, for NG (α_{p+s}), CPG (α_p), PPG (α_s), and DPG (α_{p-s}). **b** Resolution improvement and the degree of polarization, γ, as a function of the optical thickness, n. A comparison is given between CPG and NG (P_1), DPG and NG (P_2), and DPG and CPG (P_3). Reprinted with permission from [2], 1998, Elsevier

This property is illustrated more clearly in Fig. 6.6b, which shows the percentage of improvement in resolution between CPG and NG (P_1), DPG and NG (P_2), and DPG and CPG (P_3). It is noted from Fig. 6.6b that the resolution improvement by using any polarization-gating method becomes weaker as the optical thickness, n, becomes larger than 12. This feature is caused by the reduced number of less scattered photons which can be detected and by the fact that at each scattering event depolarization is weak due to the large anisotropy value, g, of the scatterers in the turbid media. The latter feature can be seen from Fig. 6.6b; photons scattered between 15 and 30 times have similar depolarization (degree of polarization, γ). Therefore, it is difficult for polarization-gating to selectively suppress the highly scattered photons. Since polarization-gating is based on the number of detectable unscattered or less scattered photons which maintain the initial degree of polarization [5], the transverse resolution is eventually degraded to that of the NG technique as the optical thickness, n, increases, which is clearly illustrated in Fig. 6.6a, b.

6.2.3 Image Resolution in a Turbid Medium of Milk

We now demonstrate the effect of polarization-gating in a more realistic tissue-like sample which consists of scatterers of different sizes. To demonstrate the effect of the polarization-gating mechanisms on image quality in tissue-like turbid media, sample4 and sample5 (see Table 6.2) are used to form images of an embedded bar. Figure 6.7 shows typical images of a 48-μm bar embedded in sample4 (Fig. 6.7a–c) and sample5 (Fig. 6.7d–f) when CPG, PPG, and DPG techniques were employed.

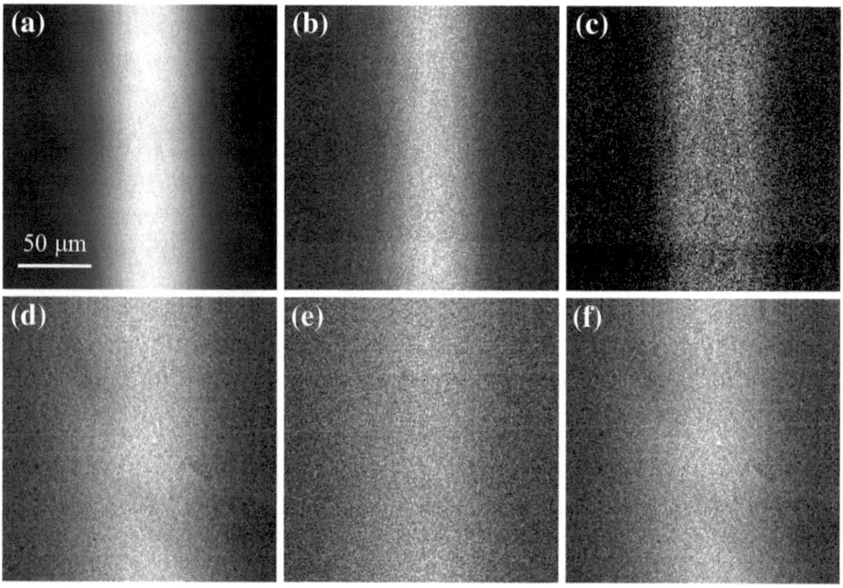

Fig. 6.7 Measured images (**a**–**c**) for sample4 (Table 6.2) and measured images (**d**–**f**) for sample5 (Table 6.2). **a** and **d**, **b** and **e**, and **c** and **f** correspond to CPG, PPG and DPG techniques, respectively. The intensity of the images is normalized to unity. The NA of the imaging objective is 0.75. Reprinted with permission from [2], 1998, Elsevier

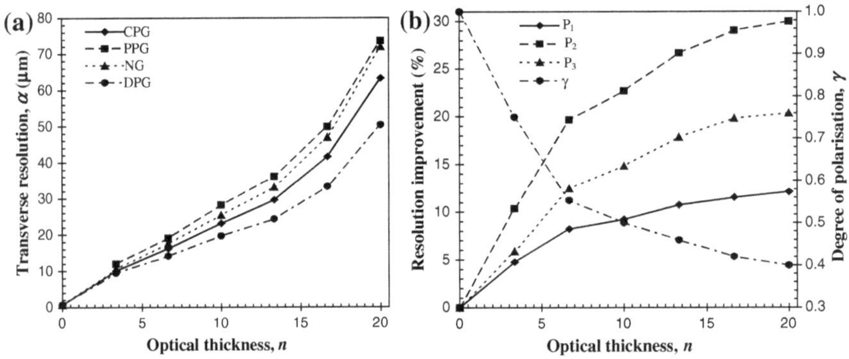

Fig. 6.8 Resolution performance for sample set E at NA = 0.75. **a** Measured transverse resolution as a function of the optical thickness, n, for NG (α_{p+s}), CPG (α_p), PPG (α_s) and DPG (α_{p-s}). **b** Resolution improvement and the degree of polarization, γ, as a function of the optical thickness, n. A comparison is given between CPG and NG (P_1), DPG and NG (P_2), and DPG and CPG (P_3)

Clearly, there is an improvement in the edge sharpness for DPG (Fig. 6.7c) when compared with CPG (Fig. 6.7a), indicated by the reduced smearing at the edges of the image. As the turbid sample becomes thicker (Fig. 6.7d–f) it is visually difficult to see the improvement in image quality between CPG (Fig. 6.7d) and DPG (Fig. 6.7f). This result suggests that at an optical thickness, n, of 20 for the semi-skimmed milk suspension, the number of less scattered photons is low. Therefore the effect of polarization-gating is diminished. From the results presented for polystyrene microspheres, the average particle size for the milk suspension may be referred to be between 0.2–0.5 μm.

The measured transverse resolution as a function of the optical thickness, n, for NG (α_{p+s}), CPG (α_p), PPG (α_s), and DPG (α_{p-s}) techniques when an imaging objective of NA 0.75 is used, is illustrated in Fig. 6.8a. It is evident from Fig. 6.8a that using CPG (α_p) is still superior than NG (α_{p+s}) in microscopic imaging. Figure 6.8a also demonstrates that DPG (α_{p-s}) is superior to CPG (α_p) in microscopic imaging. Image resolution using this technique is 10 times better than that achieved in transillumination imaging for a similar turbid medium [5]. This improvement is illustrated more clearly in Fig. 6.8b, which shows the percentage improvement in resolution between CPG and NG (P_1), DPG and NG (P_2), and DPG and CPG (P_3).

It should be noted that the improvement in resolution between DPG and CPG (P_3) is more pronounced when the optical thickness, n, is larger (>10). In this range, the amount of unscattered light detected is significantly reduced, and furthermore, the depolarization of scattered photons becomes pronounced, as is illustrated in the degree of polarization, γ, in Fig. 6.8b. It can be seen that even at an optical thickness of $n = 20$ unscattered or less scattered photons can be extracted using the DPG method, although the number of these photons are low. In other words, high-sensitivity detectors are needed to produce an image of high resolution without further signal processing.

6.2.4 Effect of Pinhole-Gating and Polarization-Gating

Polarization-gating can be used in conjunction with spatial-gating and temporal-gating techniques since these gating methods are based on different aspects of randomization of the incident light. An example of this principle is illustrated in Sects. 5.2 and 5.3, which involved using the polarization-gating mechanism to further suppress unwanted scattered photons in the angle-gating mechanism. It is also predicted in Sect. 5.4 that a finite-sized pinhole placed in front of the detector (i.e. weak confocal arrangement) can effectively reduce the contribution of scattered photons which have propagation paths not intersecting at the entrance aperture [3, 12–15].

In this section, the effect of additional spatial filtering on image resolution is experimentally investigated for sample set C with scatterers of diameter 0.48 μm, when a pinhole of 150 μm (in diameter) is placed in front of the detectors D_1 and D_2 shown in Fig. 6.1. It should be noted that the maximum optical thickness for this set of results is limited to 20, due to the reduced signal-to-noise ratio for the detected signal when pinholes are employed.

The measured transverse resolution as a function of the optical thickness, n, for NG (α_{p+s}), CPG (α_p), PPG (α_s), and DPG (α_{p-s}) techniques, is illustrated in Fig. 6.9a when a pair of 150-μm pinholes are used. Comparing Figs. 6.6a and 6.9a demonstrates that using the additional pinhole filtering significantly increases the transverse resolution for all polarization-gating mechanisms. The transverse resolution improvement for polarization-gating with and without pinholes is defined as

$$P_4\big|_n = \frac{\alpha_\varsigma\big|_n - \alpha_{\varsigma+\text{pinhole}}\big|_n}{\alpha_\varsigma\big|_n} \tag{6.4}$$

where ς represents the polarization-gating mechanisms used.

At an optical thickness, n, of 20, the transverse resolution without the pinholes (see Fig. 6.6a) is, 47.2, 52.1, 49.9, 42.1 μm for CPG, perpendicular polarization-gating, NG and differential polarization-gating, respectively. With the addition of the 150-μm pinholes, the transverse resolution is improved to (see Fig. 6.9a) 35.1 μm (or $P_4 \approx 26$ %), 38.8 μm (or $P_4 \approx 26$ %), 37.5 μm (or $P_4 \approx 25$ %), and 32.0 μm (or $P_4 \approx 24$ %), for CPG, perpendicular-polarization-gating, NG and DPG, respectively. This result demonstrates that scattered photons which are not suppressed in polarization-gating may be removed by a pinhole in terms of their path deviation.

Figure 6.9b illustrates the percentage improvement in resolution between CPG and NG, P_1, DPG and NG, P_2, and DPG and CPG, P_3, when the pinholes are used. A comparison of Figs. 6.6b and 6.9b illustrates that the difference between the transverse resolution (i.e. resolution improvement) for all polarization-gating methods becomes lower when a pinhole is used. This phenomenon is due to the efficient reduction in the number of collected multiply scattered photons which have similar polarization state, with the use of the pinhole. Therefore, when the

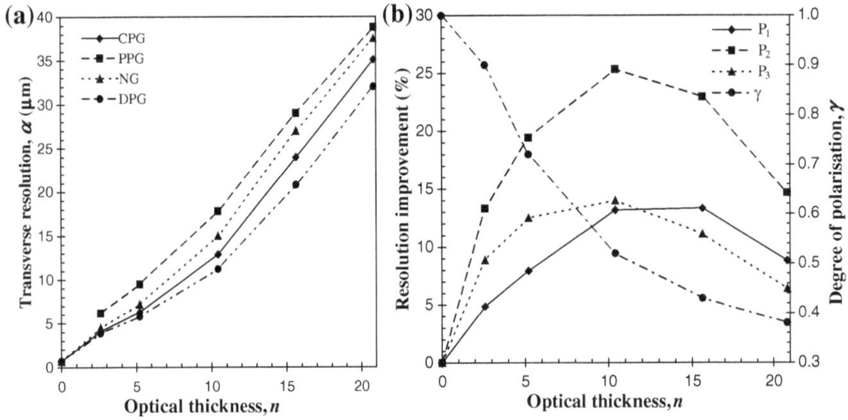

Fig. 6.9 Resolution performance for sample set C at NA = 0.75 and with a pinhole of diameter 150 μm. **a** Measured transverse resolution as a function of the optical thickness, n, for NG (α_{p+s}), CPG (α_p), PPG (α_s) and DPG (α_{p-s}). **b** Resolution improvement and the degree of polarization, γ, as a function of the optical thickness, n. A comparison is given between CPG and NG (P_1), DPG and NG (P_2) and DPG and CPG (P_3)

polarization-gating methods are applied the resolution improvement is not large. The reduction in the number of collected highly scattered photons is also illustrated by a 19 % (optical thickness, $n \sim 20$) increase in the degree of polarization, γ (see Figs. 6.6b and 6.9b) when the pinhole is employed.

6.2.5 Dependence of the Degree of Polarization on Scatter Size

It is seen from Figs. 6.3c, d, 6.5b and 6.6b that the degree of polarization, γ, approaches a non-zero background even when $n > 20$. Even in a region where all photons are scattered and are totally depolarized, for example, sample set D illustrated in Fig. 6.10, the non-zero asymptotic value of the degree of polarization, γ, is approximately 0.24 at an optical thickness, n, of 55. The reason for this phenomenon can be explained as follows.

For a reflection scanning microscope with large area detectors, the detected signal always includes a contribution from surfaces that are not in the focal volume of the imaging objective. In the case of the sample cells used in this experiment, this part of the contribution is primarily due to reflections from the glass cell surfaces. Since the incident light is polarized, the reflected light from the glass cell surfaces has almost the same polarization direction as the incident beam and only contributes an added offset to the parallel polarization intensity, I_p. Thus the degree of polarization, γ, may be artificially increased, which results in the shifting of the

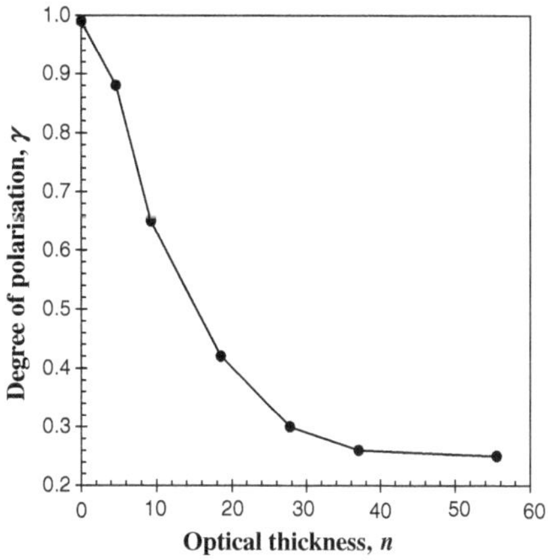

Fig. 6.10 Degree of polarization, γ, for sample set D as a function of the optical thickness, n. The NA of the imaging objective is 0.75. Reprinted with permission from [2], 1998, Elsevier

asymptotic value of the degree of polarization, γ, from zero to a value which depends on the NA of the imaging objective and the strength of the scattering signal (Figs. 6.5b and 6.6b). It should be noted that the effect of the added intensity contribution on the asymptotic value of the degree of polarization, γ, becomes more pronounced, when the optical thickness, n, is large and the corresponding intensity originating from the sample is weak. Since this added signal contribution results only in an intensity offset it does not affect the transverse resolution, α, achievable. This background contribution can be removed in image processing or reduced to negligible amounts by anti-reflection coating on all surfaces of the glass cell.

The effectiveness of DPG in suppressing diffusing photons in microscopic imaging depends on (a) the number of unscattered photons and (b) the degree of polarization, γ, of the scattered photons. Three working regimes have been introduced according to these two conditions [5], and are characterized as follows: Regime 1: unscattered light is present and scattered light retains polarization, regime 2: unscattered light is present and scattered light is depolarized or there is no unscattered light but scattered light retains polarization, and regime 3: there is no unscattered light and scattered light depolarized. Each turbid sample used in Tables 6.1 and 6.2 experiences 2 or 3 regimes as the optical thickness, n, is increased. For sample sets A and B, regimes 1 and 2 are experienced. In these two cases, scattered photons depolarize fast but there is a significant proportion of unscattered photons. In these regimes, conventional polarization-gating is effective but not as effective as DPG as illustrated in Figs. 6.3 and 6.5. For samples sets C, D, and E, all three regimes are experienced as the optical thickness, n, is increased. In other words, scattered photons depolarize slower and the number of unscattered or less scattered photons is significantly reduced and eventually they are no longer detectable (Figs. 6.6 and 6.10). Both CPG and DPG are thus less effective in regime 3 ($n > 20$) and play no role in improving image resolution as illustrated in Fig. 6.6.

6.2.6 *Effect of Numerical Aperture*

The goal of microscopic imaging is to obtain high resolution with a help of a high NA objective. Although a high NA objective gives a narrow diffraction spot at its focus, and therefore gives a high-resolution image if scattered photons are negligible, there is a problem when a high NA objective is used to image through a turbid medium. Scattered photons propagating at higher angles through such an objective result in longer path lengths that the photons traverse, which effectively increases the average number of scattering events, *n*, experienced by the photons. Therefore, the number of photons maintaining the initial degree of polarization, γ, can be reduced if a high NA objective is used (see Fig. 5.17). A lower NA objective rejects more scattered photons traveling at higher angles and as a result, image resolution is higher. Therefore, for a high NA objective, the degradation of image resolution caused by scattering photons is severe if no other gating mechanism is used. This effect is further demonstrated in Fig. 6.3a, b, which shows that an objective of NA = 0.25 gives higher resolution than an objective of NA = 0.75 when the optical thickness, *n*, is small and the depolarization for each scattering event is strong. A comparison of the resolution improvement between DPG and CPG reveals that

$$P_3|_{\mathrm{NA}=0.25} > P_3|_{\mathrm{NA}=0.75}.$$

Thus, using a smaller NA objective in differential polarization-gating leads to a stronger improvement.

Another example of demonstrating the effect of the NA on DPG is depicted in Figs. 6.11 and 6.12, in which sample6 (see Table 6.2) is imaged. Figures 6.11 and 6.12 show typical images of an embedded 48-μm bar for conventional polarization gating (Figs. 6.11a and 6.12a), PPG (Figs. 6.11b and 6.12b) and DPG (Figs. 6.11c

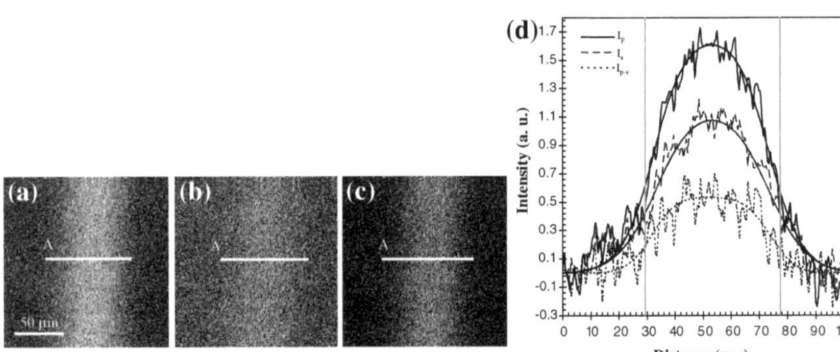

Fig. 6.11 Measured images for sample6 (Table 6.2) for CPG (**a**), PPG (**b**) and DPG (**c**) techniques. The intensity of the images is normalized to unity. **d** Includes the measured edge responses corresponding to CPG (I_p), PPG (I_s), and DPG (I_{p-s}) techniques. The *vertical gray line* represents the ideal edge position. The NA of the imaging objective is 0.25

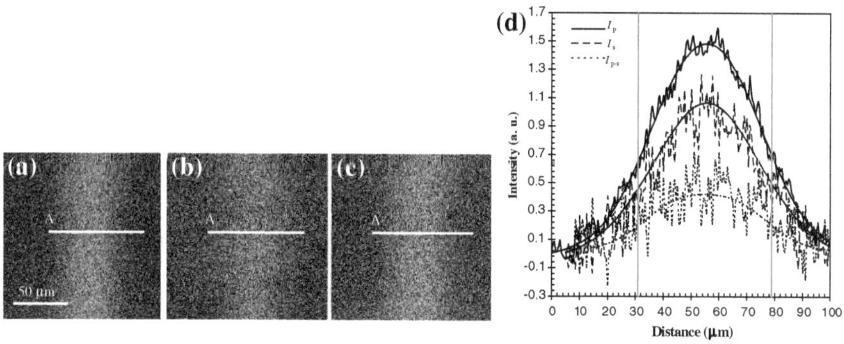

Fig. 6.12 Measured images for sample6 for CPG (**a**), PPG (**b**), and DPG (**c**) techniques. The intensity of the images is normalized to unity. **d** Includes the measured edge responses corresponding to CPG (I_p), PPG (I_s), and DPG (I_{p-s}) techniques. The *vertical gray line* represents the ideal edge position. The NA of the imaging objective is 0.75. Reprinted with permission from [2], 1998, Elsevier

and 6.12c) mechanisms for NA = 0.25 and NA = 0.75, respectively. Figures 6.11d and 6.12d give corresponding cross-sections and fitted SDS curves. In this example, the depolarization for each scattering event is weak because the anisotropy value, g, is 0.92, resulting in the degree of polarization, γ, of 0.52 and 0.41 for NA = 0.25 and NA = 0.75, respectively. The measured transverse resolution for NA = 0.25 (Fig. 6.11d) is 46.6, 52.1 and 36.6 μm for CPG (α_p), PPG (α_s), and DPG (α_{p-s}), respectively. The measured transverse resolution of an embedded bar for NA = 0.75 (Fig. 6.12d) is 61.9, 64.4, and 47.5 μm, for CPG (α_p), PPG (α_s), and DPG (α_{p-s}), respectively. It is clearly seen that using an objective of NA 0.25 gives higher transverse resolution than using an objective of NA 0.75 for all polarization-gating mechanisms. In particular, when DPG (I_{p-s}) is used, the overall shape of the transverse edge response is closer to that of the embedded bar, indicated by the flat top in Fig. 6.11d. This property results from the suppression of scattered photons by DPG. Unlike Fig. 6.3a, b, it is found that the resolution improvement between DPG and CPG for a high NA objective is stronger than that for a low NA objective, i.e.,

$$P_3|_{\mathrm{NA}=0.25} < P_3|_{\mathrm{NA}=0.75}.$$

It should be also noted that the signal-to-noise ratio (SNR) of the edge responses for the objective of NA 0.25 (Fig. 6.11d) (in this chapter SNR is quantified by the ratio of the peak-to-peak signal level to the peak-to-peak noise level) is clearly higher than that for the objective of NA 0.75 (Fig. 6.12d). This property is another indication that highly scattered photons are suppressed more efficiently in the former case. Due to the random noise associated with the scattering process, the edge response by PPG (I_s) has a lower SNR than that by CPG (I_p) since more scattered photons are collected in the former case. Therefore, even though the transverse resolution for differential polarization-gating (α_{p-s}) is higher than conventional polarization-gating (α_p), the differential polarization-gated signal (I_{p-s}) has a lower SNR, since the SNR for the differential polarization-gated signal (I_{p-s}) is mainly determined by that of the signal by PPG (I_s). However, this effect can be minimized if an average process is taken over many sample traces for signals obtained by CPG and PPG, before the differential polarization-gated signal is calculated.

6.3 Monte Carlo Simulation in a Reflection Optical Microscope

Based on the Monte Carlo simulation method detailed in Sects. 3.1–3.3, one can simulate some of the observed features presented in Sect. 6.2. Consider a turbid slab [3], illuminated by an objective (see Fig. 3.3b) of NA 0.25. The turbid medium consists of polystyrene beads ($n = 1.59$) suspended in water ($n = 1.33$). Three types of polystyrene beads of diameters 0.202, 0.48 and 1.056 μm are used as scattering particles in the turbid medium, respectively. This range of the bead size is close to that in biological tissue. The anisotropy values g (scattering mean free path l_s) corresponding to 0.202, 0.48 and 1.056 μm polystyrene beads are given in Table 2. 1 for a He–Ne laser at wavelength 0.633 μm.

6.3.1 Degree of Polarization

Under these conditions, the degree of polarization as a function of the optical thickness n which is the sample thickness d divided by the scattering mean free path length l_s, is shown in Fig. 6.13. It is noticed that for a turbid medium consisting of small scattering particles, the depolarization of the scattered light is much faster than that for large scattering particles. For example, if the light propagates through a

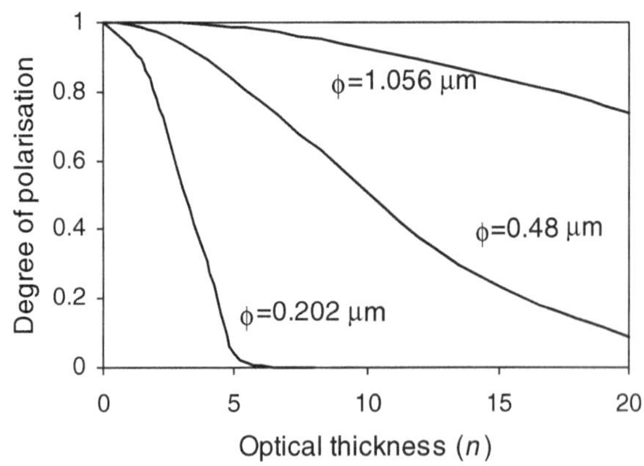

Fig. 6.13 Degree of polarization γ as a function of the optical thickness n (NA = 0.25, $v_d \to \infty$). Reprinted with permission from [3], 1999, Optical Society of America

sample of optical thickness $n = 5$, the degree of polarization drops to 4 and 99 % for a turbid medium consisting of 0.202 μm beads and 1.056 μm beads, respectively.

It should be pointed out that for a given value of n $(n \neq 0)$ the value of the degree of polarization is contributed from unscattered photons and scattered photons. Even for scattered photons which are scattered the same times, the degree of polarization is different because their propagation paths are different.

The speed of depolarization of scattered photons is related to the anisotropy value g, which reflects a directional change of photon propagation after each scattering event. For small particles, the directional change of a scattered photon is large at each scattering event due to the small value of g, which results in a significant change in the polarization state of the scattered light. For example, in the case of a turbid medium consisting of 0.202 μm beads, the light signal collected by a detector is almost totally depolarized ($\gamma \rightarrow 0$) for $n > 5$. Thus the differential polarization-gated signal approaches zero for a thick sample. On the other hand, the speed of depolarization of scattered photons for a turbid medium consisting of 1.056 μm beads is quite slow because of the large anisotropy value and thus differential polarization gating may not be efficient. For these reasons, image simulation in this section is performed only for 0.48 μm beads.

6.3.2 *Image Resolution with Polarization-Gating Methods*

In order to characterize image resolution, we assume that a high reflective edge object is embedded in a scattering slab scanned in the x direction (Fig. 3.3b). From the image intensity of the sharp reflective edge, the transverse resolution, Γ, is defined as the distance between the 90 and 10 % intensity points.

The transverse resolution for the edge object as a function of the optical thickness n is illustrated in Fig. 6.14a for different polarization-gating methods without pinhole involved. Γ_p and Γ_s are, respectively, the transverse resolution obtained with the analyser parallel and perpendicular to the polariser. Γ_{p-s} and Γ_{p+s} are, respectively, the transverse resolution obtained with the differential polarization-gating method and without any polarization-gating method. It is noticed from Fig. 6.14a that for a given NA of the objectives (NA = 0.25), Γ_{p-s} is better than Γ_p and that the differential polarization-gating method offers the highest resolution among all. The improvement becomes pronounced when n is large. This feature can be understood as follows. First since unscattered photons have a higher degree of polarization than multiply scattered photons (Fig. 6.13), more contribution from these photons can be ensured in forming an image through the use of the polarization-gating methods. Second, as explained above, scattered photons have different values of the degree of polarization for a given optical thickness n. The weighting of the scattered photons which have a lower degree of polarization can be reduced by utilizing polarization-gating methods. These two processes lead to higher resolution when the parallel and differential polarization-gating methods are used (Fig. 6.14b). However, the number of unscattered photons can be negligible in

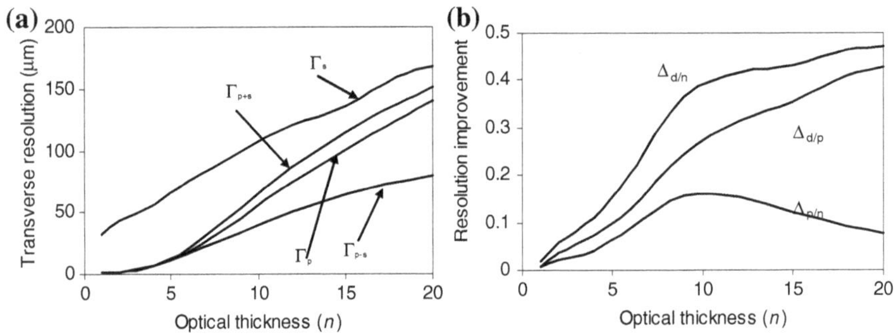

Fig. 6.14 **a** Image resolution as a function of the optical thickness n for an absorption edge embedded in a turbid medium consisting of 0.48 μm beads (NA = 0.25, $v_d \rightarrow \infty$). **b** Improvement in transverse resolution as a function of the optical thickness n for an absorption edge embedded in a turbid medium consisting of 0.48 μm beads (NA = 0.25, $v_d \rightarrow \infty$). Reprinted with permission from [3], 1999, Optical Society of America

the total photons collected when $n > 4$. In that region, the contribution from a large number of scattered photons which have a higher degree of polarization is dominant and thus the improvement in resolution becomes pronounced.

In Fig. 6.14b, the improvement in image resolution is illustrated as a function of the optical thickness, where $\Delta_{d/n}=(\Gamma_{p+s}-\Gamma_{p-s})/\Gamma_{p+s}$, $\Delta_{d/p}=(\Gamma_p-\Gamma_{p-s})/\Gamma_p$, and $\Delta_{p/n}=(\Gamma_p-\Gamma_{p+s})/\Gamma_{p+s}$. $\Delta_{d/n}$, $\Delta_{d/p}$, and $\Delta_{p/n}$ are, respectively, the improvements in image resolution between the differential polarization-gating method and the no polarization-gating method, between the differential polarization-gating method and the parallel polarization-gating method, between the parallel polarization-gating method and the no polarization-gating method. It is seen from Fig. 6.14 that as the optical thickness of a turbid medium increases, $\Delta_{d/n}$, $\Delta_{d/p}$ and $\Delta_{p/n}$ first increase and then start to saturate. There are two effects that determine the saturation of the resolution improvement. For a thicker turbid medium, on the one hand, the degree of polarization of the collected light becomes lower and more highly scattered photons can be removed by using polarization-gating methods. On the other hand, the difference of the transverse resolution between the two orthogonal states of polarization becomes less pronounced. The balance of the two effects leads to the saturation of the resolution improvement. If the turbid medium is too thick, i.e., n is too large, the resolution improvement approaches zero as the second effect becomes dominant.

Figure 6.14b shows that the resolution improvement of the differential polarization-gating method, $\Delta_{d/n}$, is approximately 47 % at $n = 20$. Because the speed of depolarization depends on the size of scattering particles (Fig. 6.13), the maximum improvement in image resolution with the differential polarization-gating method for a thick turbid medium (e.g. $n = 20$) including particles of a diameter larger or smaller than 0.48 μm may be smaller than 47 %. For example, the range of the particle size in milk suspension is from 0.1 to 1 μm and the resolution improvement

$\Delta_{d/n}$ measured for an edge embedded in a milk suspension is only 30 % (see Sect. 6.2.3).

In Fig. 6.15a, the transverse resolution as a function of the optical thickness n is shown for an edge object embedded in a turbid medium, when the diameter of the pinhole is set to v_d = 150 μm. Compared with Fig. 6.14a, in which case no pinhole is used, the transverse resolution in Fig. 6.15a is improved significantly. This improvement is because the polarization-gating methods and the pinhole-gating method are based on different aspects of randomization of the incident light. The pinhole gating method is based on the directional change of a scattered photon while polarization-gating methods are based on the polarization change due to scattering. Thus multiply scattered photons which sneak through a pinhole can be removed by polarization-gating methods, leading to a further improvement in resolution.

However, it is noticed that the difference between Γ_{p+s}, Γ_p, Γ_s and Γ_{p-s} becomes less significant, when a pinhole is used. This behavior is clearly shown in the improvement in image resolution as a function of the optical thickness (Fig. 6.15b). For example, at n = 20, the resolution improvement $\Delta_{d/n}$ drops from 47 to 23 % if a 150 μm pinhole is used. This feature can be explained as follows. It is found from the simulation that the intensity I_s of the perpendicular polarization component is reduced greatly by using the pinhole. In other words, a pinhole has already removed a large amount of multiply scattered photons which have a lower degree of polarization, and thus leads to the increase of the degree of polarization (Fig. 6.16a). Therefore when the polarization-gating methods are applied, the resolution improvement is not large.

The Monte Carlo simulation predictions shown in Figs. 6.14a, 6.15a, and 6.16a are confirmed by the experimental results obtained used the method detailed in Sect. 6.2. The measured degree of polarization as a function of the optical thickness is illustrated in Fig. 6.16b. As expected, when a pinhole is used, the degree of

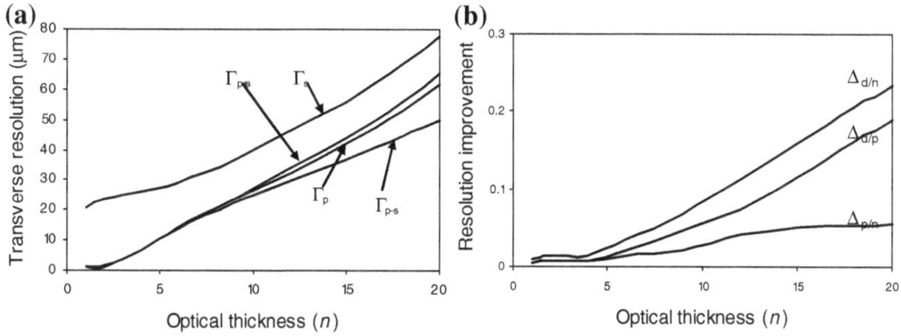

Fig. 6.15 **a** Transverse resolution as a function of the optical thickness n for an absorption edge embedded in a turbid medium consisting of 0.48 μm beads (NA = 0.25, v_d = 100 μm). **b** Improvement in transverse resolution as a function of the optical thickness n for an absorption edge embedded in a turbid medium consisting of 0.48 μm beads (NA = 0.25, v_d = 100 μm). Reprinted with permission from [3], 1999, Optical Society of America

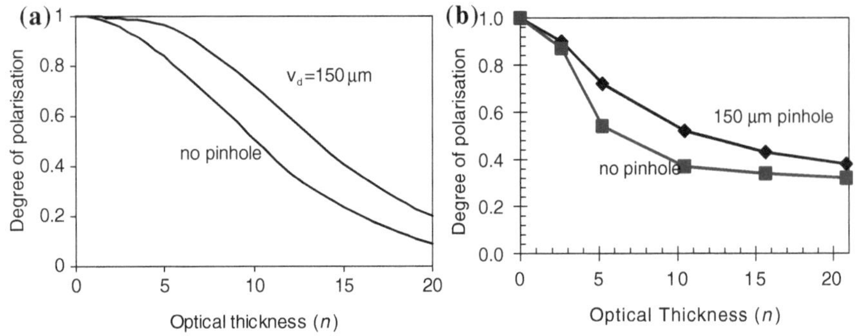

Fig. 6.16 **a** Simulated degree of polarization for a turbid medium consisting of 0.48 μm beads with and without a confocal pinhole of 100 μm. **b** Measured degree of polarization for a turbid medium consisting of 0.48 μm beads with and without a confocal pinhole of 150 μm. Reprinted with permission from [3], 1999, Optical Society of America

polarization increases, which confirms the theoretical prediction in Fig. 6.4a. The degree of polarization in the experiments is slightly higher than the theoretical prediction when n is large, because there is 4 % reflection from the glass surface of the cell used to contain turbid medium. Figures 6.6a and 6.9a qualitatively confirms that (i) using polarization-gating methods, in particular using the differential gating method, improves transverse resolution, (ii) the combination of polarization-gating and pinhole gating methods improves the image resolution further, and (iii) the resolution improvement without pinhole gating is larger than that with pinhole gating. Although Figs. 6.6a and 6.9a are obtained by an objective of NA = 0.25, these features held for NA = 0.25 [3].

6.3.3 Trade-off Between Signal Strength and Image Resolution

Polarization gating and pinhole gating can play an independent role in microscopic imaging through turbid media. One may ask which one is more efficient. Another issue related to differential polarization gating is that signal strength may be reduced significantly for a thick turbid medium. To understand these problems, we developed the criterion in Sect. 5.4 [13], which states that the efficiency of an optical gating method can be determined by the resolution at a given value of signal strength is used to compare the pinhole gating and polarization-gating methods.

The curves in Fig. 6.17 represent the relationship between transverse resolution and signal strength detected by a confocal pinhole of different sizes. Each curve terminates at point when a pinhole of v_d = 4,000 μm is used for detection, which is approximately 1,000 times larger than the airy spot of an lens on the detection plane. The signal strength here is defined as the number of detected photons divided by the total number of the incident photons propagating into the turbid medium.

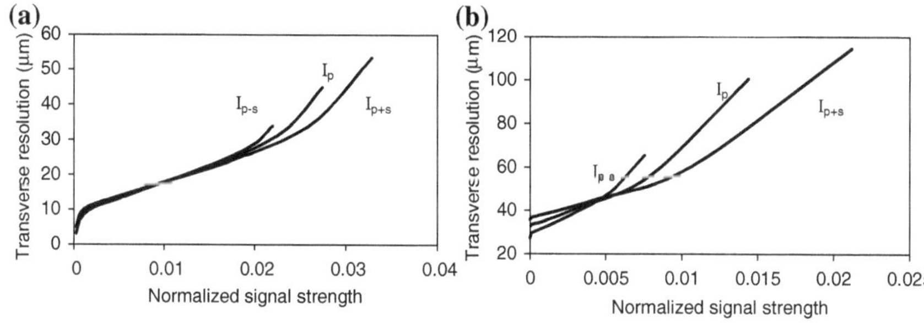

Fig. 6.17 Transverse resolution as a function of signal strength for a turbid medium consisting of 0.48 μm beads (NA = 0.25). **a** $n = 9$; **b** $n = 15$. Reprinted with permission from [3], 1999, Optical Society of America

In Fig. 6.17a, the relationship between signal strength and transverse resolution is shown for differential polarization-gating, parallel polarization gating, and no polarization gating when $n = 9$. In this situation, the degree of polarization is 62 % for v_d = 4,000 μm, which means that the signal strength detected with parallel polarization gating is significantly stronger than that detected with the perpendicular polarization gating. As a result, the difference of the curves between no polarization gating and parallel polarization gating is not pronounced even for v_d = 4,000 μm.

It is noticed from Fig. 6.17a that the transverse resolution can be improved at the expense of signal strength. When the detected signal is strong, corresponding to a large confocal pinhole, the situation without any polarization-gating methods gives highest image resolution at a given signal strength. This phenomenon means that using a pinhole only to reduce the number of scattered photons is better than the combination of polarization gating and pinhole gating methods. However, when the detected signal is weak, i.e., when a small confocal pinhole is used, the curves representing parallel and differential polarization-gating methods give a better trade-off between the signal strength and the transverse resolution. In particular, the differential polarization-gating method gives the best trade-off among all. In other words, instead of using a small pinhole, a larger size pinhole can be used together with polarization-gating methods to maintain resolution and in the meantime to retain higher signal strength to overcome the noise. Another feature in Fig. 6.17a is that the gradient of the curves changes quickly near the origin where a small pinhole (5 μm) is used. In this situation, the detected signal is weak because of the removal of a large portion of scattered photons. Therefore, the weighting of the contribution from ballistic light becomes significant, and leads to a significant improvement in image resolution which causes the change of the gradient.

In Fig. 6.5b, a turbid sample of optical thickness $n = 15$ is chosen, in which case the degree of polarization is only 23 % for v_d = 4,000 μm. Comparing the curves in Fig. 6.5b with those shown in Fig. 6.5a, we can see that the difference of the curves between polarization gating and no polarization-gating methods are more significant in the latter situation when the pinhole size increase. This feature can be expected from the fact that a lower degree of polarization leads to more efficient polarization-gating methods. As a result, the advantage of using polarization-gating methods, particularly using the differential polarization-gating method, becomes more significant even when signal strength is weak.

6.4 Monte Carlo Simulation in a Transmission Optical Microscope

In this section, with the use of the Monte Carlo simulation model (see Sects. 3.1–3.3), we simulate image resolution in a transmission optical microscope illustrated in Fig. 3.3a for turbid sample set C (see Table 6.2).

Figure 6.18a demonstrates the transverse resolution as a function of the optical thickness, n, for NG (α_{p+s}), CPG (α_p), PPG (α_s), and DPG (α_{p-s}). Figure 6.18b demonstrates the resolution improvement between CPG and NG (P_1), DPG, and NG (P_2) and DPG and CPG (P_3) when an infinitely large detector is employed. Figure 6.19a demonstrates the theoretically improved transverse resolution as a function of the optical thickness, n, with the introduction of a 100 μm pinhole, for NG (α_{p+s}), CPG (α_p), PPG (α_s) and DPG (α_{p-s}). Figure 6.19b demonstrates the resolution improvement between CPG and NG (P_1), DPG and NG (P_2), and DPG and CPG (P_3) when a 100 μm pinhole is placed in front of the detector.

Figure 6.19 theoretically confirms that a finite-sized pinhole can significantly improve the transverse resolution, α, achievable when employed in conjunction with polarization-gating and that the resolution improvement achieved with polarization-gating is reduced with the introduction of the spatial pinhole.

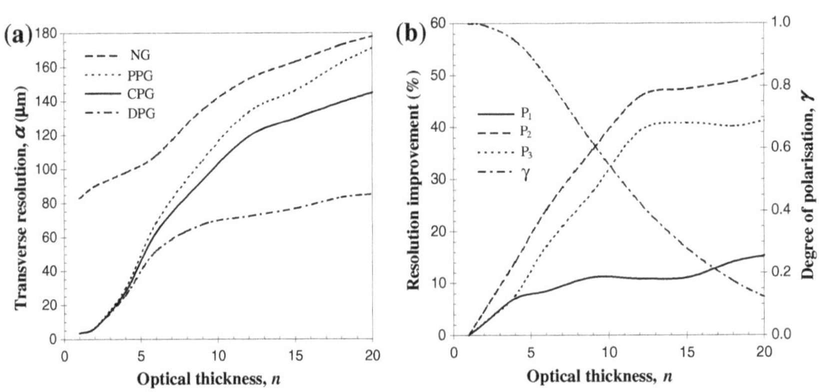

Fig. 6.18 Resolution performance for sample set C at NA = 0.25. **a** Calculated transverse resolution as a function of the optical thickness, n, for NG (α_{p+s}), CPG (α_p), PPG (α_s), and DPG (α_{p-s}). **b** Resolution improvement and the degree of polarization, γ, as a function of the optical thickness, n. A comparison is given between CPG and NG (P_1), DPG and NG (P_2), and DPG and CPG (P_3)

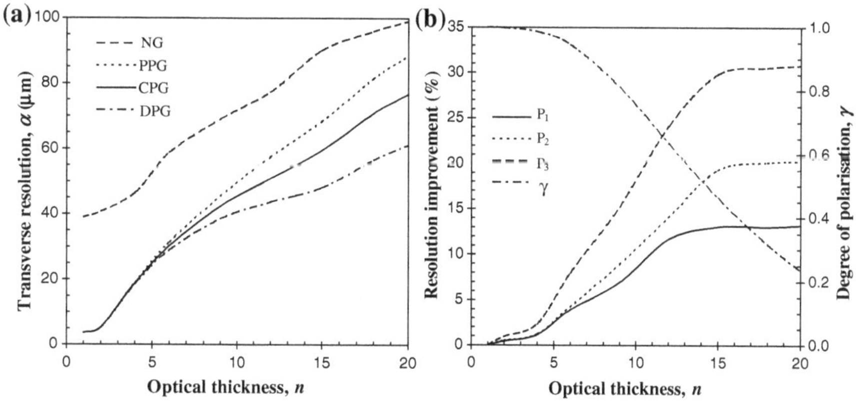

Fig. 6.19 Resolution performance for sample set C at NA = 0.25 and a pinhole of diameter 100 μm. **a** Calculated transverse resolution as a function of the optical thickness, n, for NG (α_{p+s}), CPG (α_p), PPG (α_s) and DPG (α_{p-s}). **b** Resolution improvement and the degree of polarization, γ, as a function of the optical thickness, n. A comparison is given between CPG and NG (P_1), DPG and NG (P_2), and DPG and CPG (P_3)

Figures 6.18b and 6.19b also confirm the additional suppression of scattered photons with a finite-sized pinhole which is indicated by an increase in the degree of polarization, γ.

6.5 Effective Point Spread Function

In Chap. 4, the concept of the effective point spread function (EPSF) has been introduced. As a result, the image of an object can be calculated by the convolution of an object function and the EPSF.

6.5.1 Effective Point Spread Function for Polarization Gating

For transmission optical microscope [16], consider a turbid slab, illuminated by an objective (Fig. 3.3a) of NA 0.25. The turbid medium consists of polystyrene beads (n = 1.59) suspended in water (n = 1.33). Two types of polystyrene beads of diameters 0.35 and 0.48 μm are used as scattering particles in the turbid medium, respectively. The anisotropy values g corresponding to 0.35 and 0.48 μm polystyrene beads are 0.72 and 0.81, respectively, for a He–Ne laser at wavelength of 0.633 μm. The scattering mean free path length l_s, is assumed to be 20 μm for all cases and the thickness of the turbid slab is considered as twice of the focal depth d. 40,000,000 illumination photons are used in the Monte Carlo simulation to

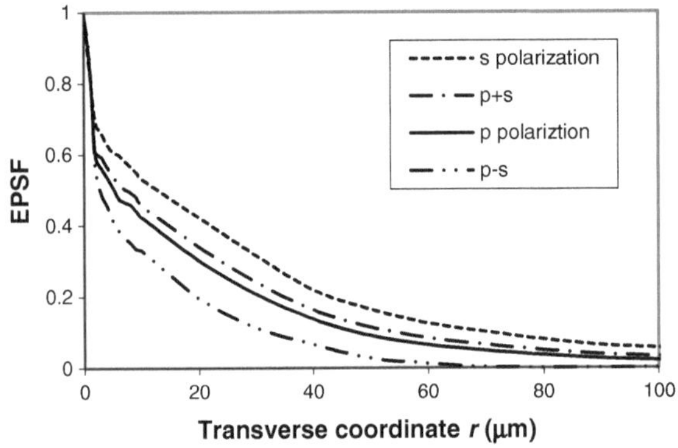

Fig. 6.20 EPSF at the focal depth d of 100 μm in a turbid medium consisting of 0.48 μm beads. $r = (x, y)$ is the transverse radial coordinate originating from the focus. Reprinted with permission from [16], 2002 SPIE

ensure the accuracy of an EPSF for a transmission-mode scanning optical microscope.

The EPSFs at the depth of 100 μm in the medium consisting of 0.48 μm beads under different polarization-gating methods are shown in Fig. 6.20. The results show an improvement of image resolution under polarization-gating methods. The improvement is not only reflected on the narrowing of the EPSF under polarization-gating methods, but also on the significant reduction of the tail of the EPSF, which indicates an efficient suppression of highly scattered photons. It should also be pointed out that the improvement is particularly significant under the differential polarization-gating method.

6.5.2 Image Resolution and Signal Level with Polarization-Gating Methods

In order to characterize image resolution, we assume that a high absorption edge object is embedded in the middle of a scattering slab scanned in the x direction. From the image intensity of the sharp absorption edge, the transverse resolution, Γ, is defined as the distance between the 90 and 10 % intensity points.

The transverse resolution as a function of the focal depth in a medium consisting of 0.48 μm beads is illustrated in Fig. 6.21a for different polarization-gating methods, Γ_p and Γ_s are, respectively, the transverse resolution obtained with the analyzer parallel and perpendicular to the polarizer. Γ_{p-s} and Γ_{p+s} are, respectively, the transverse resolution obtained with the differential polarization-gating method and without any polarization-gating method. It is shown that Γ_{p-s} is better than Γ_p and that the differential polarization-gating method offers the highest resolution among all. This feature can be understood from the difference in the degree of

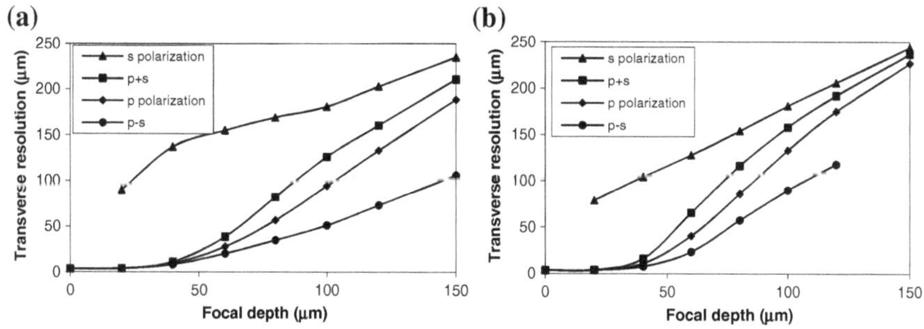

Fig. 6.21 **a** Image resolution as a function of the focal depth d for an edge embedded in a turbid medium consisting of 0.48 μm beads. **b** Image resolution as a function of the focal depth d for an edge embedded in a turbid medium consisting of 0.35 μm beads. Reprinted with permission from [16], 2002 SPIE

polarization between unscattered and scattered photons. Since unscattered and/or less scattered photons have a higher degree of polarization than multiply scattered photons, more contribution from these photons can be ensured in forming an image through the use of the polarization-gating methods. The weighting of the scattered photons can be reduced accordingly by utilizing polarization-gating methods, which leads to high resolution when the parallel and differential polarization-gating methods are used (Fig. 6.21a). It is noticed from Fig. 6.21a that the difference of transverse resolution between Γ_p and Γ_s, under two polarization states p and s, decreases with increasing the focus depth d. When $\gamma \rightarrow 0$, the two curves representing Γ_p and Γ_s approach toward each other. Because of the difference of the transverse resolution between the two orthogonal states of polarization becomes less pronounced, the resolution improvement by utilizing polarization-gating methods becomes less effective.

The transverse resolution as a function of the focal depth in a medium consisting of 0.35 μm beads is illustrated in Fig. 6.21b. The improvement of the transverse resolution under polarization-gating methods is similar to what has been demonstrated in a turbid medium with large scatterers (Fig. 6.21a). However, it is noticed that the differential polarization-gating method is only valid up to the focal depth of 120 μm. Because of the fast depolarization speed for small scatterers, the collected signal at the focal depth larger than 120 μm becomes totally unpolarized. Therefore, no differential image can be measured beyond this focal depth.

Figure 6.22a shows the dependence of the measured signal level on the sample thickness d in a medium consisting of 0.48 μm beads under different polarization-gating methods. The signal level η has been normalized by the signal measured without the turbid medium (d = 0) and any polarization-gating methods. It is noticed that the reduction in the signal level under parallel polarization gating is insignificant. However, the reduction in the signal level under differential polarization becomes significant when $\gamma \rightarrow 0$, which indicates the collect light becomes totally unpolarized. The signal level as a function of the focal depth d for a turbid

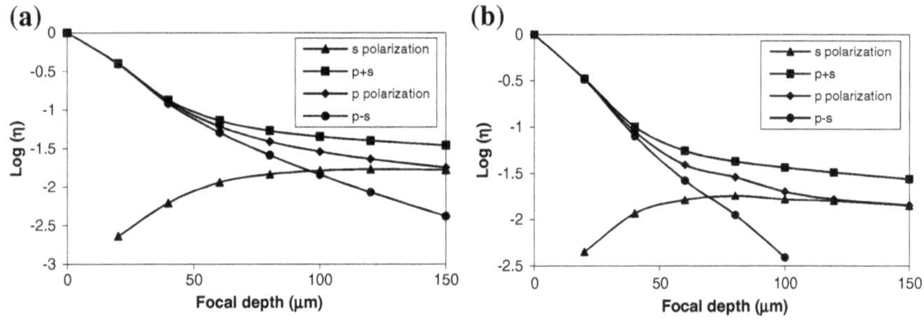

Fig. 6.22 **a** Signal level as a function of the focal depth *d* in a turbid medium consisting of 0.48 μm beads. **b** Signal level as a function of the focal depth *d* in a turbid medium consisting of 0.35 μm beads. Reprinted with permission from [16], 2002 SPIE

medium with small scatterers (0.35 μm beads) is illustrated in Fig. 6.22b. The comparison between Fig. 6.22a, b shows that at a given focal depth, the signal strength, especially the signal strength under the differential polarization-gating method, is lower in a turbid medium with small scatterers. For example, at the focal depth of 100 μm, the signal strength is five times higher in a turbid medium consisting of 0.48 μm beads than that in a turbid medium consisting of 0.35 μm beads. This is due to the faster depolarization effect in a turbid medium with smaller scatterers.

References

1. S.P. Schilders, X.S. Gan, M. Gu, Resolution improvement in microscopic imaging through turbid media based on differential polarisation. Appl. Opt. **37**, 4300 (1998)
2. S.P. Schilders, X. Gan, M. Gu, Effect of scatterer size on microscopic imaging in turbid media based on differential polarization gating. Opt. Commun. **157**, 238 (1998)
3. X. Gan, S. Schilders, M. Gu, Image enhancement through turbid media under a microscope using polarization gating methods. J. Opt. Soc. Am. A **16**, 2177 (1999)
4. S.P. Morgan, R.K. Appel, M.G. Somekh, Experimental studies of the factors affecting spatial resolution in continuous wave transillumination through scattering media (medical application). Bioimaging **2**, 163 (1994)
5. S.P. Morgan, M.P. Khong, M.G. Somekh, Effects of polarization state and scatterer concentration on optical imaging through scattering media. Appl. Opt. **36**, 1560 (1997)
6. J.M. Schmitt, S.H. Hiang, Cross-polarised backscatter in optical coherence tomography of biological tissue. Opt. Lett. **23**, 1060 (1998)
7. S.G. Demos, W.B. Wang, R.R. Alfano, Imaging objects hidden in scattering media with fluorescence polarisation preservation of contrast agents. Appl. Opt. **37**, 792 (1998)
8. A.H. Hielscher, J.R. Mourant, I.J. Bigio, Influence of particle size and concentration on the diffuse backscattering of polarised light from tissue phantoms and biological cell suspensions. Appl. Opt. **36**, 125 (1997)
9. R.G. Jensen, *Handbook of Milk Composition* (Academic Press, San Diego, 1995)
10. X. Gan, M. Gu, Temporal, angular and spatial distribution of photon migration through a highly scattering medium. Optik **108**, 129 (1998)

11. M. Gu, *Principles of Three-dimensional Imaging in Confocal Microscopes* (World Scientific, Singapore, 1996)
12. M. Gu, T. Tannous, C.J.R. Sheppard, Effect of an annular pupil on confocal imaging through highly scattering media. Opt. Lett. **21**, 312 (1996)
13. X. Gan, S. Schilders, M. Gu, Image formation in a turbid medium under a microscope. J. Opt. Soc. Am. A **15**, 2052 (1998)
14. J.M. Schmitt, A. Knuttel, M. Yadlowsky, Confocal microscopy in turbid media. J. Opt. Soc. Am. A **11**, 2226 (1994)
15. J.M. Schmitt, K. Ben-Letaief, Efficient Monte Carlo simulation of confocal microscopy in biological tissue. J. Opt. Soc. Am. A **13**, 952 (1996)
16. X. Gan, M. Gu, Image reconstruction through turbid media under a transmission-model microscope. J. Biomedical Opt. **7**, 372 (2002)

Chapter 7
Fluorescence-Gating Mechanism

Abstract The gating methods discussed in Chaps. 5 and 6 are based on the use of different optical devices or elements. Fluorescence-gating is based on the fact the wavelength of the fluorescence signal, which comes from a specific part within a turbid medium, is different from that of the excitation and the scattered signal if scattering is elastic. Therefore, selecting fluorescence signal is an effective way of removing the effect of multiple scattering in a turbid medium. In this chapter, single-photon (1p) excited fluorescence microscopy through turbid media is studied, in which, one incident photon is absorbed to excite an electron to an excited state (Pawley, Handbook of biological confocal microscopy. Plenum, New York, 1995; Gu, Principles of three-dimensional imaging in confocal microscopes. World Scientific, Singapore, 1996). We will focus the Monte Carlo simulation in this chapter, while experimental investigation will be given in Chap. 8 together with comparisons with two-photon (2p) and three-photon (3p) fluorescence microscopy. In Sect. 7.1, we define resolution and signal level since these two parameters are critical for imaging through a thick turbid medium. In Sects. 7.2 and 7.3, we study the image quality in a single-layer homogeneous and inhomogeneous turbid media, respectively. Double-layer turbid media in which a fluorescent object is embedded is investigated in Sect. 7.4. The effect of spherical scattering aggregates on the focusing of an objective through a turbid medium is presented in Sect. 7.5.

7.1 Transverse Resolution and Signal Level

Two parameters used in Chaps. 5 and 6 for evaluating the imaging performance of a microscope in a turbid medium are transverse resolution (Γ) and signal level (η). To characterize image resolution, we consider a thin fluorescent sharp edge embedded in a turbid slab of thickness d (Fig. 7.1a). The edge is modeled at a given depth f_d where the geometry focus of an imaging objective is located [1, 2]. From the fluorescence image intensity of the sharp edge scanned in the x direction, transverse resolution Γ is defined as the distance between the 90 and 10 % intensity points (Fig. 7.1b).

M. Gu et al., *Microscopic Imaging Through Turbid Media*,
Biological and Medical Physics, Biomedical Engineering,
DOI 10.1007/978-3-662-46397-0_7

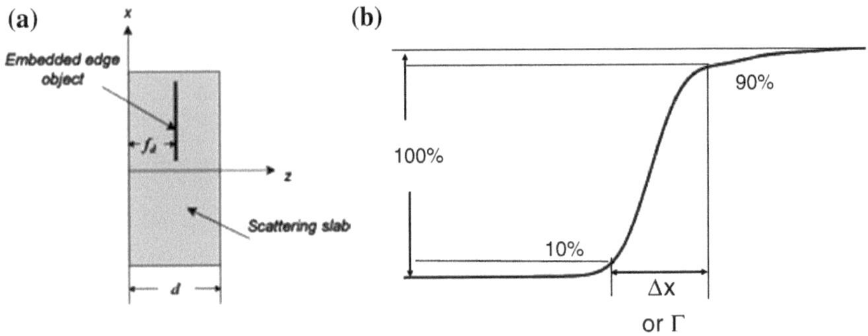

Fig. 7.1 Schematic diagram of modelling the image of a fluorescent edge and the definition of transverse resolution

The signal level η is defined as the number of fluorescence photons generated from a fluorescent sheet and collected by a detector. Usually, the signal level is normalized by the fluorescence signal strength when no scattering exists. This normalization process eliminates the difference of the strength of the excitation process, and thus allows a unique comparison of the relative decay rate of the fluorescence strength along the depth of a turbid medium. According to the definition of the signal level, the signal level represents the total contribution of fluorescence photons collected by the detector which contains the fluorescence photons excited by ballistic photons, snake photons, and scattered photons.

This definition of the signal level does not provide the direct information of the image resolution; however, it is important to determine the achievable resolution of the reconstructed image in imaging processing through a turbid medium. In this process, the object information is extracted through a deconvolution operation based on the concept of the effective point spread function (EPSF) introduced in Chap. 4 (see Chap. 9); the more the object information is extracted, the higher the image resolution. Though ballistic and snake photons carry the object information of high resolution, scattered photons also carry the object information but just in a complicated way. To achieve a stable deconvolution operation, an input image with strong strength is necessary. Otherwise, the noise signal from sources other than a turbid sample can lead to pronounced artifacts in a deconvolved image. Therefore, these parameters, the image resolution Γ and the signal level η, provide the complemental information of the image performance in a turbid medium.

The microscope considered in this chapter is a reflection system (see Fig. 3.3b) in which L_1 and L_2 are two identical objectives used for illumination and detection, respectively, and a finite-sized pinhole of diameter v_d is placed in front of the detector. The detail of the Monte Carlo simulation is given in Sect. 3.9. The imaging modeling is based on the convolution of the EPSF and a fluorescent object, as introduced in Chap. 4. Totally, 10,000,000 illumination photons are used in the Monte Carlo simulation to ensure the accuracy of an EPSF for a reflection-mode scanning optical microscope. It is assumed that under single photon excitation, the wavelength of the excitation beam and fluorescence beam is 400 nm.

7.2 Image Performance Through Single-Layer Homogeneous Turbid Media

One way to improve the image quality in imaging through a turbid medium is to suppress the contribution of highly scattered photons by using optical gating elements. As discussed in Chaps. 5 and 6, one of the most efficient optical gating devices used in microscopic imaging is the pinhole gating method. However, before a pinhole mask starts to play its role, photons emerging from a turbid medium go through a preselection process determined by the aperture of objectives. In this section, we will demonstrate the influence of the numerical aperture (NA) of objectives, pinhole and scatter size on 1p fluorescence image quality in a single homogeneous turbid layer [3].

The scattering medium consists of either spherical particles of diameter (ρ) 0.48 μm or spherical particles of diameter 0.202 μm, suspended in water. We assume that the particle concentration in the turbid medium consisting of 0.48 μm particles is $0.87 \times 10^9/\text{mm}^3$. According to Mie Scattering theory (Chap. 2), the corresponding scattering mean free path length (SMFPL) is 3.68 and 15 μm for wavelengths 400 and 800 nm, respectively. The optical thickness *n* is defined as the sample thickness *d* divided by the SMFPL. For example, if a sample is embedded at a depth of 30 μm, the corresponding optical thickness is 8.14 for wavelength 400 nm. In order to demonstrate the effect of particle size, the SMFPL in a turbid medium consisting of 0.202 μm particles is also assumed to be 3.68 and 15 μm, respectively, for wavelength 400 and 800 nm. The parameters associated with the two turbid media are summarized in Table 7.1.

7.2.1 *Effect of the Numerical Aperture of an Objective*

The EPSF at a depth of 30 μm for 1p fluorescence imaging in a turbid medium consisting of 0.48 μm scatterers is shown in Fig. 7.2a for different values of the numerical aperture of an objective. It is found that a small central peak becomes visible at the focal center for high NA objectives (NA = 1.25). The central peak is

Table 7.1 Parameters associated with scattering media

	$\lambda = 400\,\text{nm}$	
ρ (μm)	g	SMFPL (μm)
0.48 (L)	0.89	3.68
0.202 (S)	0.69	3.68

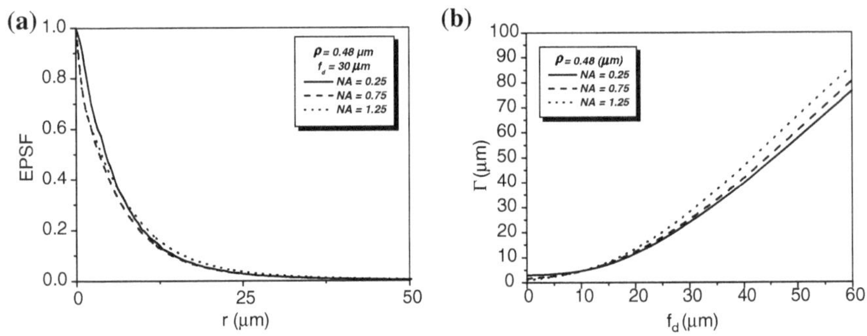

Fig. 7.2 EPSF at a depth of 30 μm (**a**) and the transverse resolution as a function of the focal depth (**b**) for different values of the numerical aperture of an objective in 1p fluorescence imaging ($v_d \rightarrow \infty$, $\rho = 0.48$ μm)

mainly contributed by ballistic photons which suffer no scattering effect on their way propagating to the focal center. These ballistic photons form a diffraction-limited spot. For high numerical aperture objectives, the focal spot is small which leads to a high intensity at the focal center. However, the other components in the EPSF, contributed by scattered photons, form a broad distribution for high NA objectives. This is because a high NA objective collects scattered photons travels at high angles and these scattered photons statistically experience more scattering events and contribute to a broader distribution in the EPSF.

In Fig. 7.2b, the corresponding transverse resolution as a function of the focal depths is illustrated for different values of the numerical aperture of an objective. It is shown that for a higher NA objective, the transverse resolution is slight poorer after a given focal depth if $f_d > 5$ μm. This is because the scattered components in the EPSF are dominant in forming an image. In this case, more scattered photons are collected by the objective, and the corresponding distribution in the EPSF is broad (see Fig. 7.2a) which leads to a poorer transverse resolution. However, a higher NA objective also gives a sharper distribution in the EPSF for ballistic component (see Fig. 7.2a), which gives a better resolution, when $f_d > 5$ μm. But the strength of such component is weaker compared with scattered component for a thicker turbid medium. As the result, using a higher NA objective without other gating methods, such as pinhole gating in 1p fluorescence imaging gives a poorer transverse resolution. However, the effect is not significant.

The normalized signal level as a function of the focal depth is demonstrated in Fig. 7.3a. It is noticed that a high NA objective results in a relatively quick degradation in the signal level as the focal depth increases. However, the relative signal level, which is the collected fluorescence photons out of the incident photons, at a given focal depth is high when a high NA objective is used (Fig. 7.3b). This is because that a high NA objective has a high collected angle thus more scattered photons away from the paths of ballistic photons are collected.

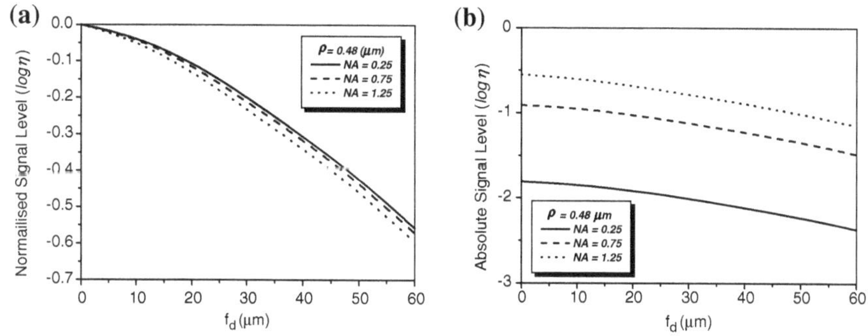

Fig. 7.3 Normalised signal level (**a**) and relative signal level (**b**) as a function of the focal depth for different values of the numerical aperture of an objective in 1p fluorescence imaging ($v_d \to \infty$, $\rho = 0.48\ \mu m$)

7.2.2 *Effect of Pinhole*

To demonstrate the effect of pinhole gating on fluorescence microscopic imaging, we show the EPSF at a depth of 30 μm for different sizes of pinhole in Fig. 7.4. It is noted that with the utilization of a confocal pinhole, the EPSF becomes slightly narrower; the smaller the size of the pinhole, the narrower the EPSF. It is also noticed that the tail of the EPSF drops quickly when a pinhole is used. This feature shows that pinhole gating is particularly efficient in suppressing highly scattered photons which deviate further from the path of ballistic photons. The elimination of the long tail may significantly reduce the blurring of an image caused by the scattering effect. In Fig. 7.5a, the transverse resolution as a function of the focal depth is illustrated for different sizes of pinhole. As expected, the transverse resolution improves when a pinhole is used. For example, at a depth of 45 μm, the transverse resolution is 62 μm when no confocal pinhole is used, but becomes 42.8 and 27.1 μm, respectively, for a pinhole of diameters $v_d = 100\ \mu m$ and $v_d = 50\ \mu m$. The ratio of improvement is 31 and 56 % in these two cases, respectively. The degradation of the signal level as the function of the focal depths is demonstrated in Fig. 7.5b. The smaller the pinhole size, the quick drop the signal level as the focal depth increases. This phenomenon can be well explained by the fact that for deeper focal depths, fluorescence phones suffer more scattering and deviate further, and consequently are suppressed by the pinhole.

7.2.3 *Effect of Scatterer Size*

Now let us turn to the effect of scatter size. The scattering medium consists of either spherical particles of diameter (ρ) 0.48 μm or spherical particles of diameter

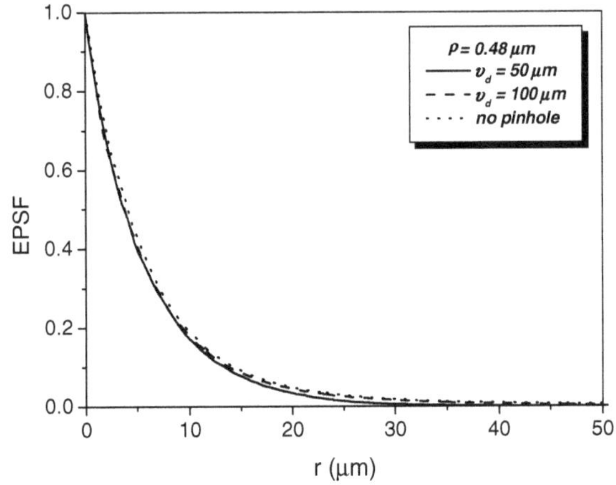

Fig. 7.4 EPSF for 1p fluorescence imaging at a depth of 30 μm for different sizes of a pinhole (NA = 0.25, ρ = 0.48 μm)

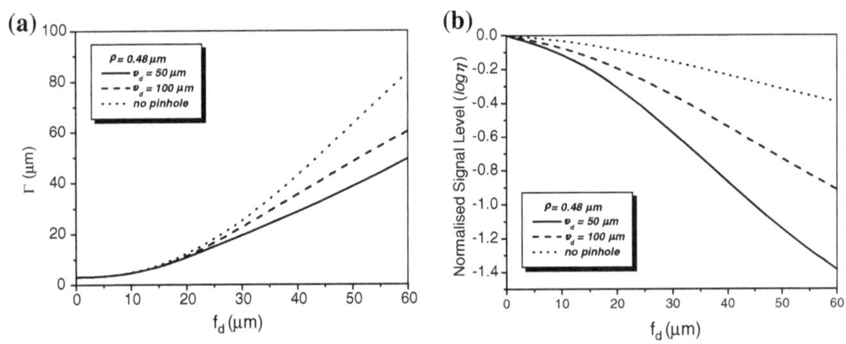

Fig. 7.5 Transverse resolution (**a**) and normalised signal level (**b**) as a function of the focal depth for different sizes of a pinhole in 1p fluorescence imaging (NA = 0.25, ρ = 0.48 μm)

0.202 μm, suspended in water. A comparison of the EPSFs for 1p fluorescence imaging of the turbid medium consisting of either 0.48 or 0.202 μm particles at depths of 10 and 30 μm is shown in Fig. 7.6. It is demonstrated that the EPSF is broader for a turbid medium consisting of small scattering particles. This feature is due to the fact that the anisotropy value g is smaller for smaller scatterers (see Table 7.1). Therefore, a scattered photon statistically deviates further from the paths of ballistic photons after each scattering event which leads to a broader EPSF. The difference of the EPSFs between large particle and small particle turbid media becomes larger as the focal depth becomes deeper.

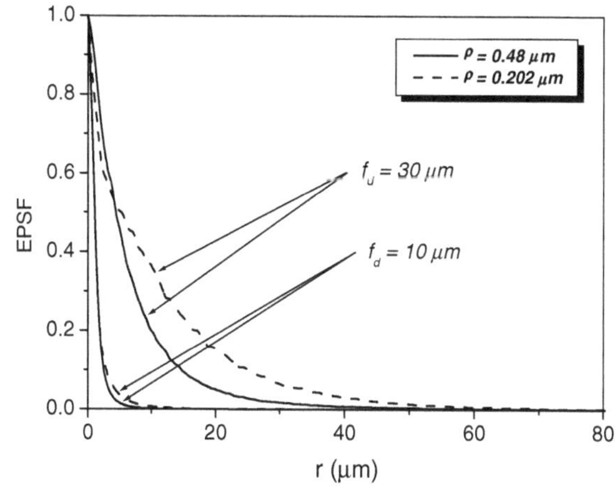

Fig. 7.6 EPSF for 1p fluorescence imaging at different depths for a turbid medium consisting of either scatterers of diameters 0.48 μm or scatterer of diameter 0.202 μm, respectively ($v_d \to \infty$, NA = 0.25)

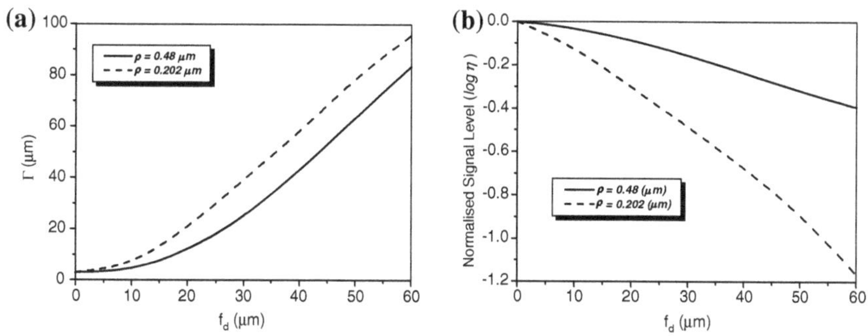

Fig. 7.7 Transverse resolution and signal level as a function of the focal depth for different sizes of scattering particles in 1p fluorescence imaging ($v_d \to \infty$, NA = 0.25)

Transverse image resolution and signal level of an edge object embedded in a turbid medium consisting of either particles of diameter 0.48 μm or particles of diameter 0.202 μm is illustrated in Fig. 7.12a, b, respectively. It is shown that the transverse resolution degrades when a turbid medium becomes thick and that the resolution becomes worse if a turbid medium consists of smaller scatterers as the result of a broader EPSF shown in Fig. 7.7a.

As the imaging depth becomes thicker, the signal level also decreases (Fig. 7.7b). It is also noticed that the degradation of the signal level is much quicker as the function of the focal depths in the turbid medium consisting of the smaller scatterers due to the smaller anisotropic value g, which results in further deviation from the collection paths.

7.3 Image Performance Through Single-Layer Inhomogeneous Turbid Media

In this section, 1p fluorescence microscopy through turbid media that consist of multisized scattering particles is investigated. It is known that the scattering features of a medium can be described by two parameters, the anisotropy value (g) and scattering cross-section (σ) of the scattering particles. In fact, the concentration of the particles of a given size in a turbid medium also plays an important role in the determination of the scattering features of the medium. Combined with the cross-section, it determines the total cross-section of the turbid medium through an effective mean free path length as introduced in Sect. 3.7. Therefore, the investigation into the size effect actually includes the effect of the size distribution and the effect of the concentration distribution [4].

7.3.1 Scattering Property of a Multi-sized Turbid Layer

According to the size distribution of small organelles and nuclei in a cell [5], three groups of inhomogeneous turbid media are considered for understanding image formation. Groups S and L, which attempt to investigate the scattering features from small organelles and large nuclei in cells, contain nine types of scattering particles. Their diameter (ρ) covers from 0.1 to 0.9 μm and from 2.6 to 3.4 μm, respectively, with a diameter interval of 0.1 μm. The mean size of these two groups is 0.5 and 3.0 μm, respectively. The third group of the turbid media (M) is a mixture of groups S and L, which can be used to simulate the situation where small scattering organelles and large nuclei both exist. Based on Mie theory, the corresponding anisotropy value (g_{i}) and scattering cross-section (σ_{si}) of each type of particles are shown in Table 7.2.

Two types of the concentration distributions, the uniform and Gaussian distribution, are also to be investigated, which are, respectively, represented by U or G. All concentration distributions discussed in this section are centrally symmetrical through the mean size particle. Each turbid medium is labeled by two letters followed by two digitals. The first letter represents the type of particle group, while the second letter denotes the type of the concentration distribution. The two digitals correspond to a width parameter δ. The width parameter δ in the medium with a uniform distribution is $\delta_{\mathrm{p}}/2(\delta_{\mathrm{p}} = \rho_{\max} - \rho_{\min})$ where $\rho_{\max}$ and $\rho_{\min}$ are the maximum and minimum particle sizes included in the medium, respectively. In the medium with a Gaussian distribution, the width parameter δ is defined as the full width at half maximum of the concentration distribution. For example, the SG01 medium contains a small group of particles with a Gaussian distribution and $\delta = 0.1$ μm.

The parameters used in the Monte Carlo simulation are as follows. 10^7 illumination photons are used to ensure the accuracy of simulation results. The numerical aperture of the objective is chosen to be 0.25. It is assumed that the wavelength of

Table 7.2 Scattering parameters (σ_s, g) of the small (S) and large (L) groups of scattering particles (Reprinted with permission from [4], 2003, SPIE)

ρ (μm) (S)		0.1	0.2	0.3	0.4	0.5	0.6	0.7	0.8	0.9
400 nm (1p excitation) (fluorescence)	σ_{si} (μm^2)	0.000296	0.0081	0.0483	0.155	0.36	0.686	1.132	1.678	2.28
	g_i	0.193	0.685	0.804	0.864	0.894	0.916	0.924	0.928	0.929
ρ (μm) (L)		2.6	2.7	2.8	2.9	3.0	3.1	3.2	3.3	3.4
400 nm (1p excitation) (fluorescence)	σ_{si} (μm^2)	14.28	15.1	15.65	16.16	16.62	16.56	16.23	15.98	16.44
	g_i	0.889	0.896	0.892	0.877	0.863	0.863	0.870	0.877	0.872

Table 7.3 Effective scattering mean free path length (l') of the investigated turbid media (Reprinted with permission from [4], 2003, SPIE)

Type of medium		l' (μm) 400 nm (1p excitation) (fluorescence)	Type of medium		l' (μm) 400 nm (1p excitation) (fluorescence)
SU	SU00/SG00	4.6296	LU	LU00/LG00	3.6029
	SU01	4.2699		LU01	3.6850
	SU02	3.7832		LU02	3.6863
	SU03	3.3224		LU03	3.7573
	SU04	2.9837		LU04	3.8000
SG	SG01	4.5736	LG	LG01	3.5956
	SG02	4.3182		LG02	3.6360
	SG03	3.9779		LG03	3.7065
	SG04	3.6824		LG04	3.6324
M	SG04 + LU02	1.8422			

the excitation and fluorescence beams, λ_{ex} and λ_{fluo} under 1p excitation is both 400 nm. To keep the consistent simulation conditions, we assume that the averaged total geometrical cross-section (σ) is constant in all media, which can be mathematically expressed as

$$\int \pi(\rho/2)^2 c(\rho)\,\mathrm{d}\rho = \pi(\rho_0/2)^2 c_0 \tag{7.1}$$

where ρ is the diameter for a particle type existing in an inhomogeneous medium and $c(\rho)$ is the corresponding concentration. Here ρ_0 refers to the scattering particle size in a homogeneous medium with concentration c_0. In our calculation, $\rho_0 = 0.5\ \mu\mathrm{m}$ and $c_0 = 0.6/\mu\mathrm{m}^3$ for the uniform S group. Accordingly, in the uniform L group, $\rho_0 = 3\ \mu\mathrm{m}$ and $c_0 = 0.0167/\mu\mathrm{m}^3$.

According to the concentration distributions and the scattering cross-section at different excitation wavelengths (Table 7.2), the effective mean free path length (l') (see Sect. 3.7) of the various inhomogeneous media can be calculated and are summarized in Table 7.3.

7.3.2 Effect of Size Distributions

In order to investigate the effect of the size distribution, all the scatterers in a medium are supposed to have the same concentration which corresponds to the uniform distribution U. As an example, EPSFs for turbid media SU00 and SU04 at different focal depths under 1p excitation are demonstrated in Fig. 7.8. The sharp peak shown in Fig. 7.8 demonstrates the existence of ballistic and less scattering

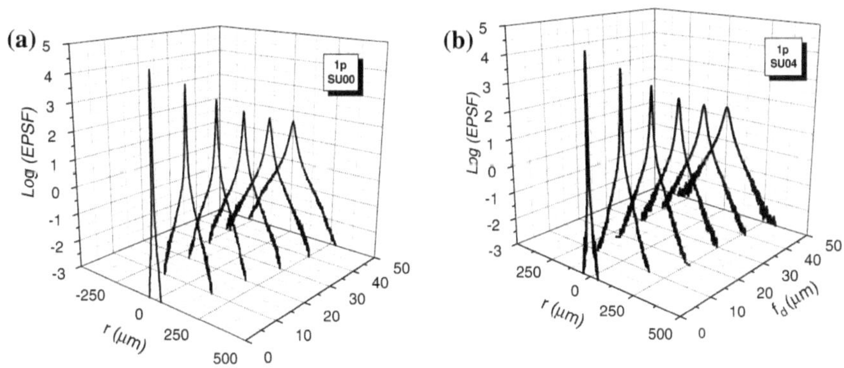

Fig. 7.8 EPSFs for fluorescence imaging at different focal depths in SU00 and SG04 turbid media. Reprinted with permission from [4], 2003, SPIE

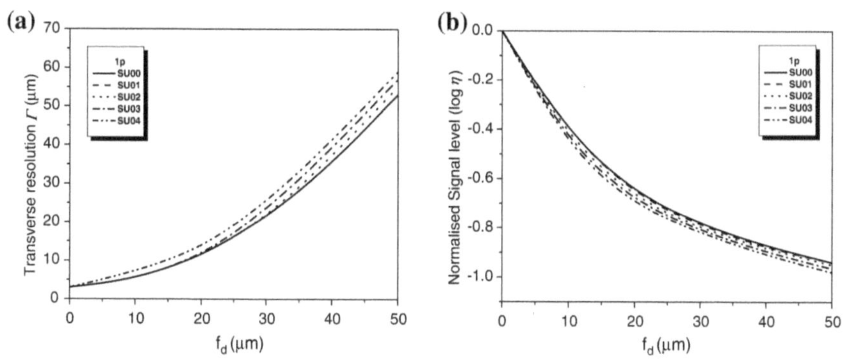

Fig. 7.9 Transverse image resolution (**a**) and signal level (**b**) as a function of the focal depth in the SU media. Reprinted with permission from [4], 2003, SPIE

snake photons. It is noticed that as the focal depth increases, the EPSF becomes broad and the peak value decreases. This is because as the focal depth increases, the scattering events increase, which leads to the broadening of the EPSF. Due to strong scattering, the number of the ballistic and snake photons that undergo fewer scattering events around the center of the focus region decreases. As a result, the peak value reduces. A comparison of the EPSFs between SU00 and SU04 media demonstrate that at a given focal depth, the EPSFs are narrower in the homogeneous SU00 media.

The transverse image resolution and signal level in a series of SU media are demonstrated in Fig. 7.9. Although the total geometrical cross-sections of the size distributed turbid media are equal, optical microscopic image formation in those media is affected differently due to the different scattering features of the scatterers of various sizes. The more types of particles around the mean size of 0.5 μm are included, the worse the image performance on resolution and signal level.

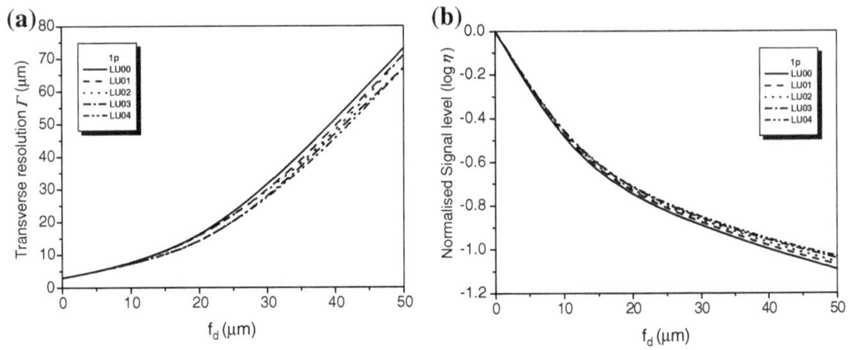

Fig. 7.10 Transverse image resolution (**a**) and signal level (**b**) as a function of the focal depth in the LU media. Reprinted with permission from [4], 2003, SPIE

However, under 1p excitation, the effect of the size distribution is not as obvious. This situation may be explained by the effective mean free path length (l') shown in Table 7.3 which clearly shows that in the medium which has the δ value of 0.4 μm (SU04), l' has the shortest value.

Figure 7.10 shows the transverse resolution and signal level in a series of the LU media. Like the situation in the SU media, the difference of the transverse resolution and signal level between different types of the media is also not significant. However, it is furthermore noticed that the LU00 medium presents the worst imaging performance rather than LU04 in particular under 1p excitation. This phenomenon is expected from the shorter l' value in Table 7.3.

7.3.3 *Effect of Concentration Distributions*

For comparison, turbid media of a Gaussian concentration distribution and of a mixed uniform and Gaussian concentration distribution are investigated.

Figure 7.11 shows the transverse resolution and signal level in the SG media. Compared with those in the SU media (Fig. 7.9), under 1p excitation, the difference of transverse resolution and signal level is not obvious. Similarly for the L group media of a Gaussian concentration distribution, the transverse resolution and signal level are shown in Fig. 7.12. There is little difference of transverse resolution and signal level between the LG and LU media under 1p excitation. Comparison of Figs. 7.9, 7.10, 7.11 and 7.12 shows that the image resolution and signal level in the media with a Guassian concentration distribution are better than those in the media with a uniform distribution. However, the impact of the size and concentration distributions in the L media is not significant.

Another comparison of Figs. 7.9, 7.10, 7.11 and 7.12 indicates that the effect of the size and concentration distributions on image performance is significant. The S and L media exhibit dramatic different features of image formation; the variation

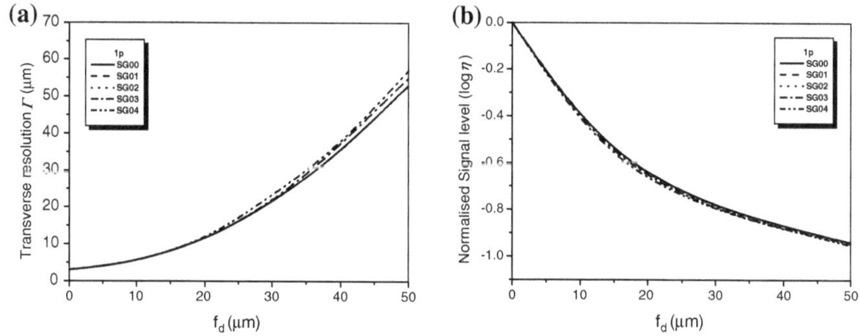

Fig. 7.11 Transverse image resolution (**a**) and signal level (**b**) as a function of the focal depth in the SG media. Reprinted with permission from [4], 2003, SPIE

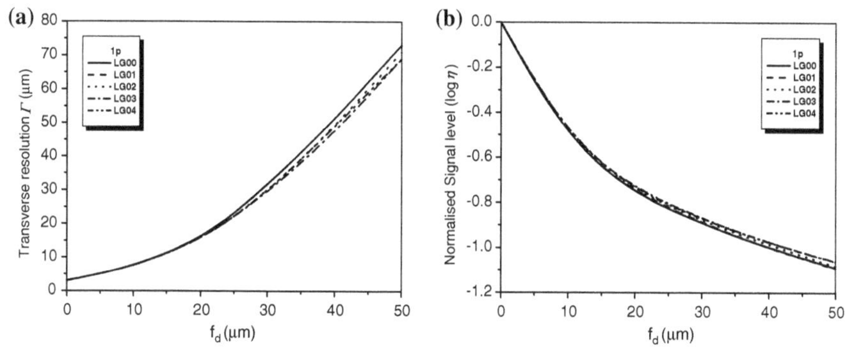

Fig. 7.12 Transverse image resolution (**a**) and signal level (**b**) as a function of the focal depth in the LG media. Reprinted with permission from [4], 2003, SPIE

of the size distribution in the L media has a slight impact on image resolution and signal level, while the corresponding influence in the S media is obvious. It is also noticed that at a given focal depth, imaging in the S media exhibits better transverse resolution and signal level than that in the L media.

These phenomena can be explained from the relationship of the scattering efficiency Q and the anisotropy g value to a scattering parameter (a/λ) (Fig. 7.13). According to Chap. 2, the scattering efficiency Q is defined as the ratio of the scattering cross-section (σ_s) to the geometrical cross-section (σ) and a is the radius size of a scattering particle. The smaller scattering efficiency implies a reduction of multiple scattering events which leads to high image resolution and signal level. In our case, the S media include particles which have relative sizes, a/λ, within the ranges from 0.125 to 1.125 under 1p excitation. While the L media include the relative particles sizes within the ranges from 3.25 to 4.25 under 1p excitation. Thus the scattering Q changes significantly as a/λ falls within the range of 0.125–1.125 in the S media and shows a slight variation while a/λ falls within

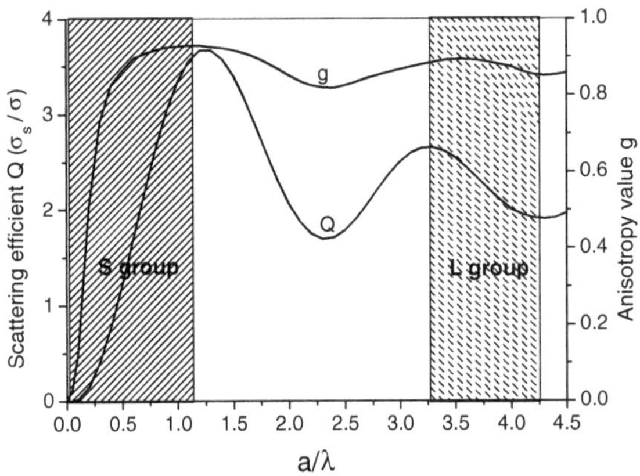

Fig. 7.13 Scattering efficiency Q and anisotropy value g as a function of the scattering parameter a/λ. The refractive indices of spherical scatterers and the immersion medium (water) are 1.59 and 1.33, respectively. Reprinted with permission from [4], 2003, SPIE

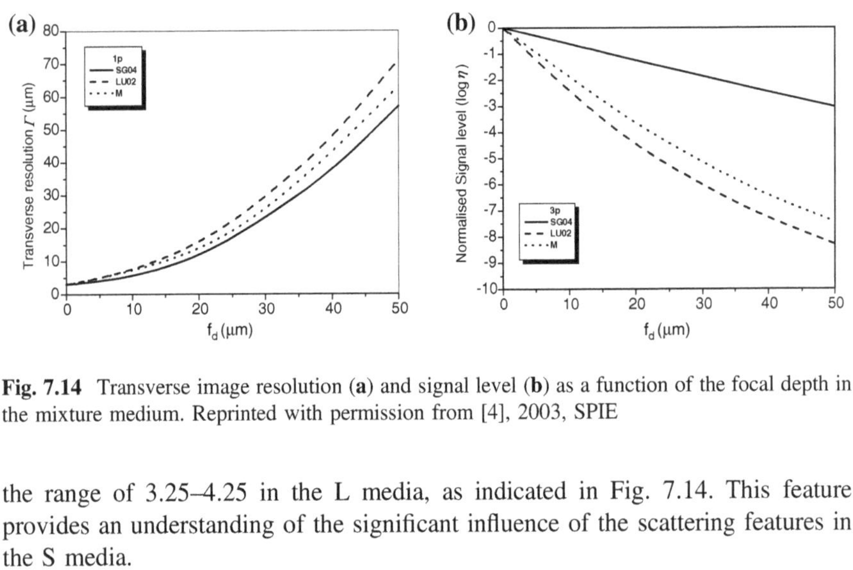

Fig. 7.14 Transverse image resolution (**a**) and signal level (**b**) as a function of the focal depth in the mixture medium. Reprinted with permission from [4], 2003, SPIE

the range of 3.25–4.25 in the L media, as indicated in Fig. 7.14. This feature provides an understanding of the significant influence of the scattering features in the S media.

Finally, let us consider a mixture medium to simulate the combined effect of the size distribution and the concentration distribution. In this simulation, it is supposed that there is a diverse distribution of particles of small sizes and a relatively concentrated distribution of particles of large sizes. This condition corresponds to the situation in cells where small scatterers (such as small organelles) distribute diversely while the size of the scattering particles like nuclei has a relatively small variation. It is noticed from Figs. 7.10 and 7.12 that for the L group, the imaging performance between the uniform and Gaussian distribution is not significant and the effect of the size distribution is not pronounced. Therefore, in the mixture medium M, we arbitrarily choose LU02 to be a simulation medium of nuclei. On the other hand, SG04 is selected to simulate the maximum effect of the size distribution of small scatterers. However, in order to keep the equality of the total geometrical cross-section, this mixture medium M has half of the concentration values for each of its corresponding constitution scatterers compared to those in the media SG04 and LU02.

Figure 7.14 demonstrates the image transverse resolution and signal level under 1p excitation in this mixture medium M. For comparison, transverse resolution and signal level in the SG04 and LU02 media are also demonstrated in the same plots. It is noticed that image performance in a mixture medium M (either transverse resolution or signal level) falls in between two extreme cases in which turbid medium only consists either small or large particles and the image performance is more affected by the LU02 medium than the SG04 medium.

7.4 Image Performance Through Double-Layer Turbid Media

So far, the turbid media considered are a single-layer structure. However, biological tissue usually exhibits a complex layer structure such as skin tissue [6]. The detail of the Monte Carlo simulation model for fluorescence microscopic imaging through double-layer turbid media has been given in Sect. 3.6. A schematic diagram of a reflection fluorescence microscope for imaging through a double-layer turbid medium is shown in Fig. 7.15. The top and bottom layers of thicknesses d_1 and d_2 are labeled with L_1 and L_2, respectively. The focal depth f_d is defined as the distance between the medium surface and the focal plane.

7.4.1 Image Performance Through Double-Layer Homogeneous Turbid Media

To compare the results of a double-layer turbid medium with those from a single-layer turbid medium, we choose two kinds of spherical scattering particles suspended in water. The first kind, called large particles (L), has a diameter ρ of 0.48 μm, and the second one, called small particles (S), has a diameter of 0.202 μm. Each layer in the double-layer structure consists of either large particles or small particles. If the top layer (L_1) includes large particles and the bottom layer (L_2) includes small particles, the double-layer turbid structure is called the LS medium. Otherwise, it is called the SL medium. The thickness of each of the two layers of the media is assumed to be 60 μm. The numerical aperture of the objective was chosen to be 0.25. It is assumed that the wavelength of the excitation beam, λ_{ex}, is 400 nm and that the fluorescence wavelength λ_f is also 400 nm. According to Mie

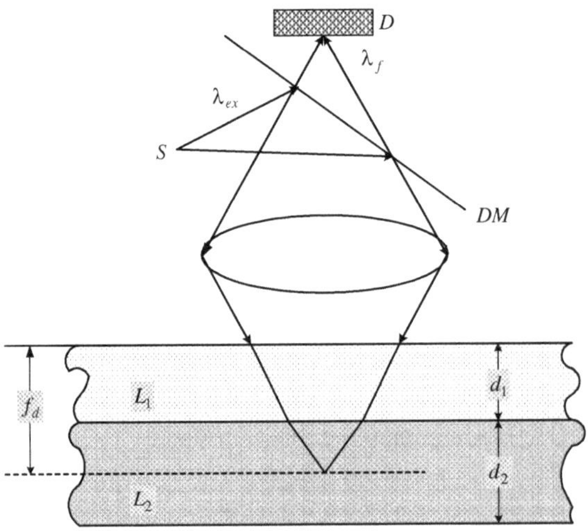

Fig. 7.15 Schematic diagram of a reflection fluorescence microscope for imaging through a double-layer turbid medium. *S* source; *D* detector; *DM* dichromic mirror; L_1 *and* L_2 *top* and *bottom* layers; d_1 *and* d_2 the thickness of L_1 and L_2; f_d the focal depth; λ_{ex} and λ_f the wavelength for excitation and fluorescence. Reprinted with permission from [7], 2002, American Institution of Physics

Scattering theory, the corresponding anisotropy value *g* and the SMFPL l_s are shown in Table 7.1.

In the Monte Carlo simulation, 10^7 illumination photons were used to ensure the accuracy of an EPSF. The EPSFs for 1p fluorescence imaging at three focal depths in the LS and SL scattering media are shown in Fig. 7.16a, b, respectively. The focal depths of 40, 60, and 80 μm mean that the focus is within the top layer, at the boundary, and within the bottom layer, respectively. As expected, when the focal depth increases, the EPSF becomes broader in both LS and SL media because of the increase of the scattering events.

A comparison of the EPSFs between the LS and SL media shows two physical features. First, at a given focal depth (Fig. 7.17), the EPSF in the LS case is narrower than that in the SL case. Second, the difference of the EPSF between LS and SL media in the bottom layer is smaller than that in the top layer. These properties can be understood from the change in the anisotropy value *g* in the LS and SL media. According to Sect. 5.4, a smaller scattering particle with a lower anisotropy value *g* results in a larger scattering angle and thus a broader distribution of scattered photons. Consequently, when the focal depth is moved into the bottom layer, the broadening of the EPSF in the SL medium becomes slower, whereas that in the LS medium becomes more quick, which leads to the behavior in Fig. 7.17b.

Based on the EPSF, we can investigate image quality in the double-layer turbid medium in terms of image resolution and signal level. The transverse image resolution and the signal level as a function of the focal depth in the LS and SL media under 1p excitation are shown in Fig. 7.18a, b, respectively. In general, as the focal

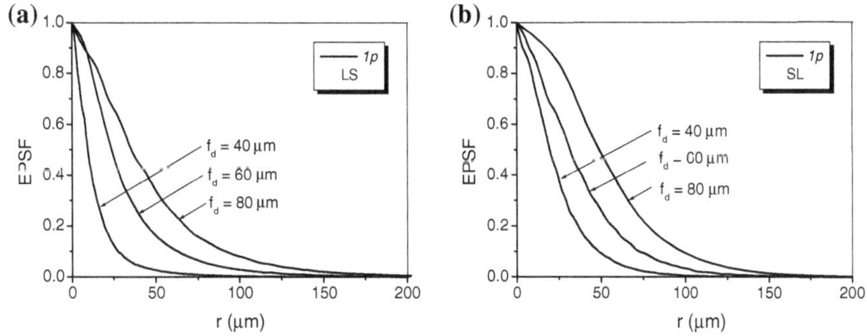

Fig. 7.16 EPSF for 1p fluorescence imaging at different focal depths in the double-layer turbid medium: **a** in the LS medium; **b** in the SL medium. Reprinted with permission from [7], 2002, American Institution of Physics

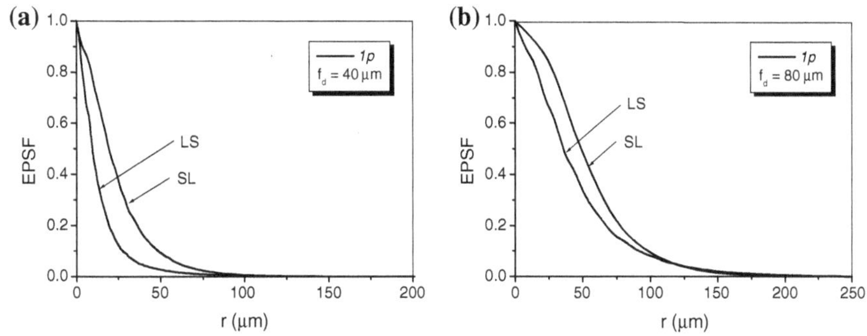

Fig. 7.17 Comparison of EPSFs for 1p fluorescence imaging between LS and SL media. **a** f_d = 40 μm; **b** f_d = 80 μm. Reprinted with permission from [7], 2002, American Institution of Physics

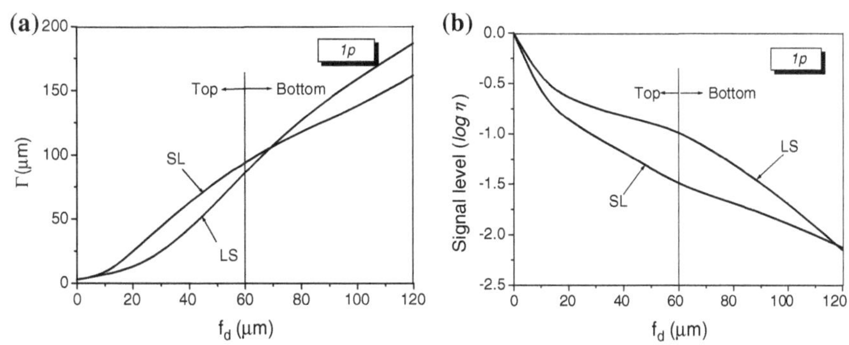

Fig. 7.18 Transverse image resolution (**a**) and signal level (**b**) as a function of the focal depth in the LS and SL media under 1p excitation. Reprinted with permission from [7], 2002, American Institution of Physics

depth increases, the transverse resolution becomes poor and the signal level is reduced. Within the top layer, the transverse resolution in the LS medium is better than that in the SL medium. This result is consistent with the behavior of the EPSF shown in Fig. 7.16.

To understand the behavior near the interface, we define the rate of resolution degradation, β, as $\beta = \mathrm{d}\Gamma/\mathrm{d}(f_\mathrm{d})$. It can be seen from Fig. 7.18a that β below the interface in the SL and LS media is slightly slower and larger than that above the interface, respectively. This feature can be explained by the change in the anisotropy value g in the two layers. In general, photons in the S layer (smaller anisotropy value g) are statistically scattered into a larger angle than those in the L layer (larger anisotropy value g). It has been demonstrated in Sect. 5.4 that the photons scattered at a larger angle lead to lower resolution than the photons scattered at a smaller angle. As a result, the increase of the anisotropy value g from the S layer to the L layer in the SL medium results in the decrease of β near the interface. In LS medium, the situation is just reserved. This feature implies that the two resolution curves may lead to a cross point at a certain focal depth in the bottom layer, as demonstrated in Fig. 7.18a.

In the top layer, the signal level in the LS medium drops more slowly than that in the SL medium, as may be expected from the anisotropy value g in the corresponding layers. The larger the anisotropy value g, the smaller the scattered angle and thus the higher the signal level. Once the focal depth is within the bottom layer, the signal level of the LS medium drops more quickly than that in the SL medium. Because the anisotropy value g in the bottom layer of the LS and SL media reduces and increases, respectively. As a result, the signal level in the SL medium is lower than that in the LS medium for the given thickness of the media.

7.4.2 Image Performance Through Double-Layer Skin Media

As a demonstration of the significance of the double-layer turbid medium model, we use it to calculate resolution and signal level in human skin tissue under 1p excitation [7]. Human skin tissue is a complex and highly scattering thick tissue. It can be considered to be a double-layer structure mainly consisting of epidermis and dermis [8]. We assume that the wavelength is 365 and 450 nm for 1p excitation and fluorescence, respectively. The absorption and scattering parameters of skin tissue at these wavelengths are summarized in Table 7.4 [6–8]. According to the

Table 7.4 Scattering parameters of the turbid media under 1p, 2p and 3p excitation

Skin layers	Epidermis			Dermis		
	μ_a (cm^{-1})	g	l_s (μm)	μ_a (cm^{-1})	g	l_s (μm)
1p Excitation (λ_{ex} = 365 nm)	100	0.72	9.1	7	0.72	21.8
Fluorescence (λ_{fluo} = 450 nm)	58	0.75	14.3	4.1	0.75	35.1

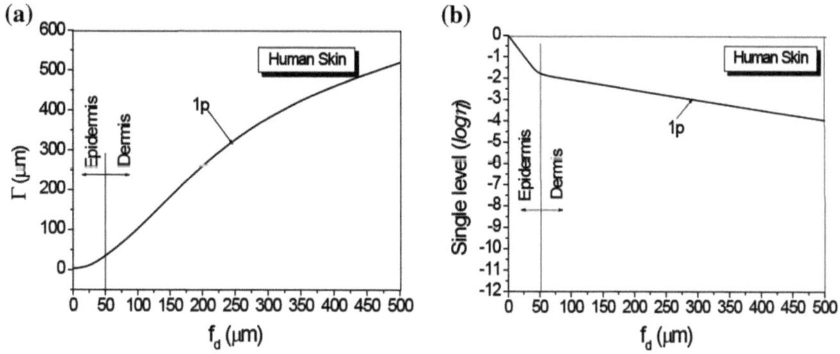

Fig. 7.19 Transverse image resolution (**a**) and signal level (**b**) in skin tissue under 1p excitation. Reprinted with permission from [7], 2002, American Institution of Physics

anatomical structure of human skin [7, 8], it can be assumed that the thickness of the epidermis and dermis layers is 50 and 450 μm, respectively.

The image resolution Γ and the signal level η of the human skin double layers under three excitation situations are shown in Fig. 7.19a, b, respectively. It is noticed that, even within the epidermis layer, the resolution is worse than that of the diffraction-limited value. The signal level in the epidermis layer (Fig. 7.19b) decays more quickly than that in the dermis layer.

7.4.3 Image Performance Through Double-Layer Human Cortex Media

Cortex is one of the most important parts of brain tissue, which is involved in almost all functions of a brain such as perception, motor coordination, thought, emotion, intellect, memory, and so on. Brain cortex, especially gray matter, is a basic research object when brain functions and diseases are studied. Imaging through brain cortex tissue can be achieved by many modalities, such as magnetic resonance imaging (MRI) and positron emission tomography (PET). However, optical microscopy offers the only way of achieving micrometer spatial resolution. In particular, fluorescence microscopy provides an effective way of investigating biological microstructures, monitoring biological activities, and diagnosing human disease.

Brain cortex has a double-layer structure that is composed of gray matter and white matter, with gray matter lying on top. Gray matter contains neurons and provides the actual processing capacity while white matter provides communications between different gray matter areas and between gray matter and the rest of tissue. Recently, scattering parameters in human brain cortex have been thoroughly investigated [9]. It is, therefore, possible and meaningful to investigate the penetration depth through human brain cortex under multiphoton microscopy, which is important for studying human brain disease in biopsy and autopsy [10].

According to the recent MRI study [11], the thickness of the gray matter layer is between 1,000 and 4,500 μm, and its average thickness is approximately 2,500 μm. Therefore, the thickness of the gray and white matter layers is assumed to be 1,000 μm for a thorough investigation of the effect of the boundary between gray matter and white matter on image performance. This human cortex structure is named Cortex. The excitation and fluorescence wavelengths are assumed to be 400 nm. Optical parameters of gray matter and white matter in human cortex [9] adopted in the simulation are summarized in Table 7.5. It should be pointed out that the assumed fluorescence wavelength does not affect the conclusion of the investigation, as the scattering parameters do not vary appreciably in the region near this wavelength [9].

The 1p fluorescence EPSFs at the focal depths of 800, 1,000, and 1,200 μm in Cortex are shown in Fig. 7.20 for an objective of numerical aperture 0.25. The focal depth f_d is defined to be the distance between the surface of the gray matter layer to the focal plane. It is clear that the focal depths of 800, 1,000, and 1,200 μm mean that the focus is within the gray matter layer, at the boundary and within the white matter layer. It is seen that when the focus moves into the white matter layer, the EPSF dramatically becomes broad. The reason for this feature is that the white matter layer has a shorter scattering mean free path length (l) compared with that in the gray matter layer, which results in a stronger scattering effect and accordingly a dominant contribution of scattered photons in image formation.

Image resolution Γ and signal level η along the entire Cortex thickness for the numerical aperture of 0.25 and 0.75 are shown in Fig. 7.21a, b, respectively. It is understandable from Fig. 7.21a that only within a shallow depth in the gray matter (f_d < 250 μm), image resolution is better than 100 μm for NA = 0.25. For signal level in Cortex (Fig. 7.21b), it is noticed that the decrease in signal level in the gray matter layer is slower than that in the white matter layer.

Table 7.5 Absorption and scattering parameters of human cortex under 1p excitation (μ_a absorption coefficient, μ_s scattering coefficient, l scattering mean free path length; g anisotropy value)

Human cortex	Optical parameters	Excitation and fluorescence wavelength 400 nm
Gray matter	μ_a (1/mm)	0.25
	μ_s (1/mm)	12.5
	l (μm)	78.5
	g	0.85
White matter	μ_a (1/mm)	0.3
	μ_s (1/mm)	42
	l (μm)	23.6
	g	0.75

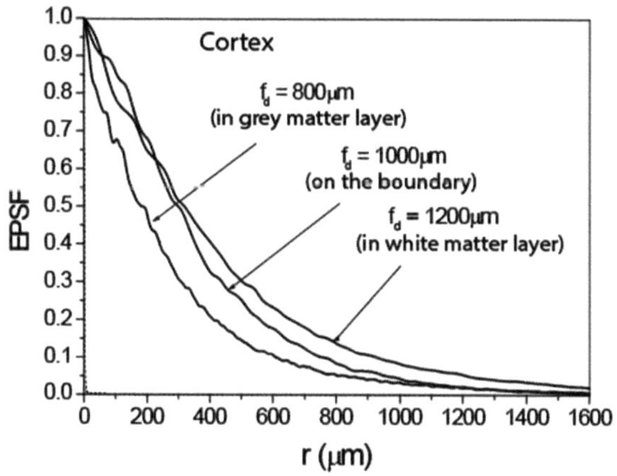

Fig. 7.20 1p fluorescence EPSFs in Cortex (numerical aperture is 0.25) at the focal depths of 800 μm (within the gray matter layer), 1000 μm (on the boundary) and 1200 μm (within the white matter layer). Reprinted with permission from [10], 2003, Optical Society of America

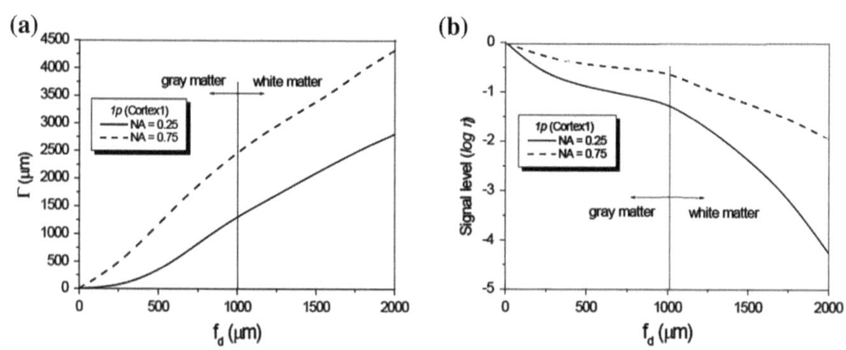

Fig. 7.21 Transverse resolution Γ (**a**) and signal level (**b**) as a function of the focal depth in Cortex for different values of the numerical aperture. Reprinted with permission from [10], 2003, Optical Society of America

It should be pointed out that an image is formed in two steps in an image system. The first step is excitation and the second is collection. It should be also pointed out that both ballistic and scattered photons in the excitation process can result in fluorescence emission. Only ballistic photons are confined to the focal region, while scattered photons in the excitation process are distributed away from the focal region. In the collection process, these two types of photons behave differently because of multiple scattering. Both processes are affected by the value of numerical aperture, as demonstrated in Fig. 7.21a, b. It is noticed that Γ is distinctly poorer as a higher NA objective is used. Beer's law indicates that scattered photons are dominant in image formation when the focal depth of the turbid medium is larger than a few scattering mean free-path length. Even when the number of scattering events is one, i.e., $f_d = 1$, approximately 64 % of the incident photons are scattered. Therefore accordingly to Table 7.5, the contribution of scattered photon is dominant in the excitation and collection processes in the case of Cortex when the focal depth is longer than 100 μm. Compared with the lower numerical aperture objective, the higher numerical aperture objective has a larger angle of convergence, and thus more scattered photons that deviate from the ballistic path may be collected by the objective. These scattering components do not satisfy the diffraction theory and make a significant contribution to image formation, leading to the degradation of the image resolution. However, the larger the NA of an objective, the slower the dropping speed in the signal level (Fig. 7.21b).

7.5 Effect of Aggregation

In Sect. 3.8, an effective Mie scattering (EMS) model has been established for the investigation into a fractal aggregate as a single scatterer. According to the Monte Carlo simulation method, the scattering parameters of a spherical aggregate, the scattering efficiency Q_a, and the anisotropy g_a value, are shown in Fig. 3.9 to reveal their dependence on the physical size D and the effective mean free path length l_m of the aggregate. The significance of the scattering efficiency Q_a and the anisotropy value g_a derived from the EMS model is that they allow for the Monte Carlo simulation of photon migration through a complex turbid medium made up of aggregates [12].

To demonstrate this, we simulate the focusing feature of an objective (NA = 0.25) through such a complex turbid medium using these two parameters. To this we use the following conditions. The excitation wavelength λ is 400 nm. The diameter of the aggregate D is 5 μm and the diameter of the scattering particles, suspended in air, which constitute the aggregate with a fractal dimension of $m = 25$ is $d = 0.2$ μm and have no absorption. The refractive index n of the scattering particles is 1.59.

Figure 7.22 shows the logarithmic representation of the focal spot of an objective as a function of the focal depth f_d in a homogeneous turbid medium of fractal aggregates. Here r is the radial distance in the focal plane. It is seen from Fig. 7.22 that as the focal depth f_d increases, the narrow central peak disappears and the focal spot becomes broad, which means that the ballistic or snake photons around the focus decrease and that the scattering photons that deviate from the center path increase. Figure 7.22 implies that image quality, such as image resolution and signal level, of an object embedded in an aggregate-distributed layered turbid medium where a fluorescent object is embedded can be studied using the methods in Sects. 7.2–7.4.

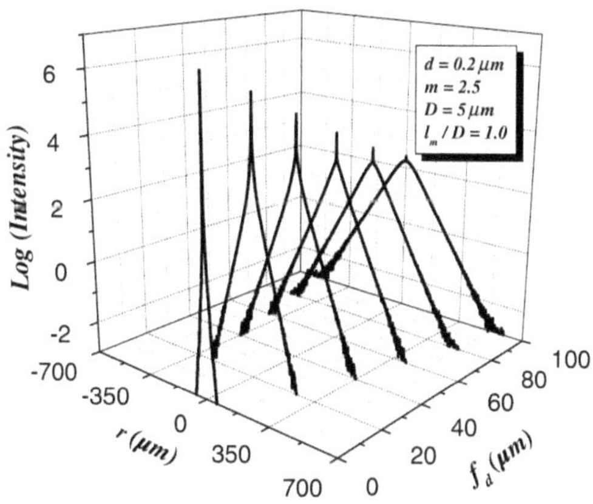

Fig. 7.22 The logarithmic representation of the excitation profiles (i.e. the focal spots) of an objective in a homogeneously random medium consisting of fractal aggregates with a concentration of $0.002/\mu m^3$. Reprinted with permission from [12], 2004, Optical Society of America

References

1. J. Pawley, *Handbook of Biological Confocal Microscopy* (Plenum, New York, 1995)
2. M. Gu, *Principles of Three-dimensional Imaging in Confocal Microscopes* (World Scientific, Singapore, 1996)
3. X. Gan, M. Gu, Fluorescence microscopic imaging through tissue-like turbid media. J. Appl. Phys. **87**, 3214 (2000)
4. X. Deng, X. Gan and M. Gu Monte-Carlo simulation of multi-photon fluorescence microscopy imaging through inhomogeneous tissue-like turbid media, *J. Biomedical Opt.* **8**, 400 (2003)
5. W. Ganong, *Review of Medical Physiology*, 16th edn. (Appleton and Lange, Norwalk, Connecticut, 1993)
6. V. Tuchin, *Tissue Optics* (SPIE Press, Bellingham, 2000)
7. X. Deng, X. Gan, M. Gu, Multi-photon fluorescence microscopic imaging through double-layered turbid tissue media. J. Appl. Phys. **91**, 4659 (2002)
8. B.R. Masters, P.T.C. So, E. Gratton, Multiphoton excitation fluorescence microscopy and spectroscopy of in vivo human skin. Biophys. J. **72**, 2405 (1997)
9. H.J. Schwarzmaier, A. Yaroslavsky, I. Yaroslavsky, T. Goldbach, T. Kahn, F. Ulrich, P.C. Schulze, R. Schober, *in Proceeding of SPIE Lasers in Surgery: Advanced Characterization, therapeutics, and System VII*, ed by R.R. Anderson, K.E. Bartels et al. Optical properties of native and coagulated human brain structures **2970**, 492 (1997)
10. X. Deng, M. Gu, Penetration depth of single-, two- and three-photon fluorescence microscopic imaging through human cortex structures: Monte-Carlo simulation, Appl. Opt. **42**, 3321 (2003)
11. B. Fischl, A.M. Dale, Measuring the thickness of the human cerebral cortex from magnetic resonance images. Proc. Natl. Acad. Sci. **97**, 11044 (2000)
12. X. Deng, X. Gan, M. Gu, Effective Mie scattering of a spherical aggregate and its application in turbid media. Appl. Opt. **43**, 2925 (2004)

Chapter 8
Multiphoton Fluorescence Imaging

Abstract As has been discussed in the previous chapter, the trade-off between image resolution and signal level plays a crucial role in determining the efficiency of optical gating methods under single-photon (1p) fluorescence imaging. Another important development in fluorescence imaging is the two-photon (2p) scanning fluorescence microscopy because of its advantages over 1p scanning fluorescence optical microscopy (Denk et al. Science 248:73–76, 1990; Schilders and Gu Appl Opt 38:720–722, 1999; Guo et al. Appl Opt 36:968–970, 1997). First, 2p excitation is a nonlinear process and therefore the fluorescence intensity is directly proportional to the square of the excitation intensity. Due to this quadratic dependence, the 2p imaging technique provides a pinpoint excitation/detection method at a deep position within thick samples. Second, if an infrared laser beam is employed for 2p excitation, 2p fluorescence microscopy offers access to ultraviolet (UV) excitation without using UV lasers, and reduces Rayleigh scattering appreciably, provided that the size of scatterers in tissue is smaller than the illumination wavelength. According to these properties of 2p excitation, it has been claimed that 2p excitation results in a deeper penetration depth than 1p excitation (Denk et al. Science 248:73–76, 1990; Guo et al. Appl Opt 36:968–970, 1997). The trade-off between image resolution and signal level for 2p fluorescence imaging is discussed in Sect. 8.1, which determines the penetration depth. The effects of various parameters on 2p fluorescence imaging are discussed in Sect. 8.2. Section 8.3 discusses the Monte Carlo simulation of 2p fluorescence imaging through various complex modeling structures. Section 8.4 presents the comparison of 1p and 2p imaging with three-photon (3p) imaging through skin and human cortex media.

8.1 Image Resolution and Signal Level

In this section, a detailed investigation into image resolution and signal level under two-photon (2p) excitation is presented [1–3]. A simulation model for developing the effective point spread function for multiphoton fluorescence imaging is proposed.

M. Gu et al., *Microscopic Imaging Through Turbid Media*,
Biological and Medical Physics, Biomedical Engineering,
DOI 10.1007/978-3-662-46397-0_8

8.1.1 Monte Carlo Simulation Model and Effective Point Spread Function

The Monte Carlo simulation model developed for 2p fluorescence imaging is similar to that for single-photon (1p) fluorescence imaging [4] described in the previous chapter. A parameter p is assigned to each photon to represent its weighting in signal strength. 1p fluorescence excitation is a linear process in which the excited fluorescence intensity is proportional to the incident intensity. Therefore, the weighting of an emitted fluorescence photon is linearly proportional to the weighting of the incident photon. However, 2p fluorescence is a nonlinear process in which the excited fluorescence light intensity has a quadratic response to the excitation light intensity:

$$I_{2\mathrm{p}} = \alpha_{2\mathrm{p}} I_{\mathrm{ex}}^2, \tag{8.1}$$

where $\alpha_{2\mathrm{p}}$ is a 2p fluorescence coefficient, and $I_{2\mathrm{p}}$ and I_{ex} are, respectively, the intensity of the incident and the fluorescence light at the focal region. Because of the nonlinear process, the implementation of (8.1) in the Monte Carlo simulation model under 2p excitation is not as straightforward as that under 1p excitation. The execution of the Monte Carlo simulation for 2p excitation is divided into two stages. In the first stage, the intensity distribution of the excitation light at the focal plane, $I_{\mathrm{ex}}(r)$, where r is the radial coordinate, is calculated and stored in the database. In the second stage, the simulation starts at the focal plane, and fluorescence photons are generated by a uniform random generator. However, the weighting factor for a fluorescence photon at a distance r is determined according to the square of the local intensity of excitation light:

$$p_{2\mathrm{p}}(r) = \alpha_{2\mathrm{p}} I_{\mathrm{ex}}^2(r). \tag{8.2}$$

When scattering is strong in a turbid medium, the illumination light distributed outside the focal region becomes more significant, which weakens the optical sectioning property of a 2p fluorescence microscope. Recent experimental research [5, 6] has demonstrated this phenomenon and an approximate theoretical model, which is based on the attenuation of ballistic light by scattering, has been developed to estimate the three-dimensional (3-D) fluorescence intensity distribution in a turbid medium under 2p excitation [7, 8]. The 3-D fluorescence intensity distribution is important to study 3-D image formation in turbid media. However, ignoring the scattered component that is usually stronger than the ballistic component may lead to significant discrepancies in predicting such a 3-D spatial distribution. To overcome this problem, we use the Monte Carlo simulation method to study 1p and 2p fluorescence imaging under a microscope with a particular interest in their 3-D fluorescence light distribution [9].

Here we use a simple approach to extend this model to study the 3-D fluorescence intensity distribution within a turbid medium. First, a stack of planes is identified within a sample volume. By recording the number of photons and their transverse locations where they pass through a given plane, a two-dimensional (2-D) transverse excitation intensity distribution $I_{ex}(x, y)$ can be obtained. Then combining all the 2-D intensity distributions from the stack of planes results in a 3-D excitation intensity distribution $I_{ex}(x, y, z)$. Unlike the model described in [9], which only considers the contribution of fluorescence excitation by ballistic light, the approach used here takes the contribution of fluorescence excitation by both ballistic and scattered photons into consideration. This treatment becomes possible because the Monte Carlo simulation method predicts the behavior of scattered photons as well as ballistic photons.

Consider a turbid medium, illuminated by an objective of numerical aperture (NA) 0.25, and assume that the geometric focal plane is 600 μm beneath the surface. The turbid medium consists of polystyrene beads (reflective index $n_i = 1.59$) suspended in a fluorescence solution such as Rhodamine 590 Tetrafluoroborate dye solution. We assume the illumination wavelength for 1p and 2p excitation is 0.532 and 1.064 μm respectively. Polystyrene beads have a diameter of 0.304 μm, and its corresponding anisotropy values g is 0.25 and 0.72, according to Mie scattering theory, for illumination wavelengths 1.064 and 0.532 μm respectively. The most significant advantage of using 2p excitation instead of 1p excitation is that the illumination wavelength used in 2p excitation is much longer, which results in a smaller scattering cross-section σ_s. For a spherical scattering particle ($n_i = 1.59$) of diameter 0.304 μm suspended in water ($n_i = 1.33$), the scattering cross-section σ_s is 0.0042 and 0.026 μm^2, respectively, for illumination wavelengths 1.064 and 0.532 μm. For a given turbid medium, the scattering-mean-free-path-length (SMFPL) under 2p excitation is more than six times longer than that under 1p excitation.

The 3-D fluorescence intensity distribution for 1p excitation $I_{1p}(x, y, z)$ is shown in Fig. 8.1. Notice that the emission of fluorescence light is limited in the focal region when scattering is weak in a turbid medium (Fig. 8.1a). However, when the turbid medium becomes denser, the amount of fluorescence light excited outside the focal region becomes more significant (Fig. 8.1b, c). It is also seen that the

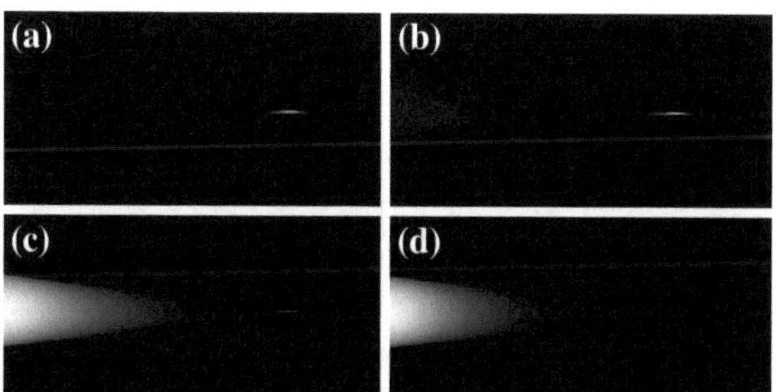

Fig. 8.1 3-D fluorescence intensity distribution $I_{1p}(x, y, z)$ for 1p excitation (Numerical Aperture = 0.25): **a** $n = 2$, **b** $n = 4$, **c** $n = 6$, **d** $n = 8$. Reprinted with permission from [9], 2000, Optical Society of American

maximum intensity point in $I_{1p}(x, y, z)$ may shift from the geometric focal region toward the sample surface if the ballistic component in the focal region is negligible (Fig. 8.1d). To explain the features demonstrated in Fig. 8.1, we need to understand the property of two components of photons, the ballistic photon component and the scattered photon component, in the propagation of an illumination beam through a turbid medium. The distribution of ballistic photons depends on the distance between the observation plane and focal plane; the closer the observation plane to the focal region, the narrower the distribution may be formed. Due to the multiple scattering effect, scattered photons deviate from the ballistic path and form a broader distribution in a turbid medium. The ballistic component is dominant in the region close to the sample surface where multiple scattering is insignificant. With the increasing depth into a turbid medium, the scattered component becomes dominant at the focal plane. However, the intensity in the focal region shows a noticeable peak if the turbid medium is not dense (Fig. 8.1a–c), due to the contribution from the ballistic component. Although the total number of ballistic photons is insignificant at a depth of 600 mm, they are concentrated in the focal center, and therefore show a noticeable peak. The ballistic peak becomes weaker when the turbid medium becomes denser, in which situation the total number of ballistic photons becomes negligible (Fig. 8.1d).

The 3-D fluorescence intensity distribution $I_{2p}(x, y, z)$ for 2p excitation is shown in Fig. 8.2. It should be pointed out that 2p excitation has a much deeper penetration in a given medium due to the fact that longer wavelength illumination light is used. In order to further demonstrate the difference between 1p and 2p excitation, we use the same set of optical thickness in Fig. 8.2. A comparison of Figs. 8.1 and 8.2 shows that for 2p excitation, the fluorescence intensity is significantly stronger in the focal region in relation to the surrounding area (Fig. 8.2a–c) until the scattering is too strong to have a non-negligible ballistic component in the focal region (Fig. 8.2d). The difference between Figs. 8.1 and 8.2 arises from the quadratic intensity dependence in 2p excitation, which efficiently enhances the fluorescence light excited by the ballistic peak in the focal region.

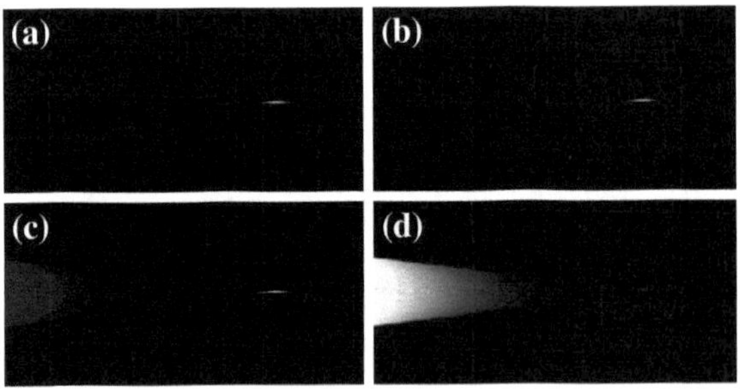

Fig. 8.2 3-D fluorescence intensity distribution $I_{2p}(x, y, z)$ for 2p excitation (NA = 0.25): **a** $n = 2$, **b** $n = 4$, **c** $n = 6$, **d** $n = 8$. Reprinted with permission from [9], 2000, Optical Society of American

To simulate the image of a fluorescence object embedded in turbid media, we introduce an EPSF under 2p excitation. The EPSF is derived from the product of the probability of an excitation photon reaching the focal plane and the probability of a fluorescence photon reaching the detector and the detail of the derivation has been discussed in Chap. 4. The significance of an EPSF is that it enables us to separate the information about an object from a surrounding turbid medium and an imaging system. It is assumed that the excitation and fluorescence wavelengths are 800 and 400 nm, respectively, under 2p excitation. The scattering medium consists of either spherical particles of diameter (ρ) 0.48 μm or spherical particles of diameter 0. 202 μm, suspended in water. We assume that the particle concentration in the turbid medium consisting of 0.48 μm particles is $0.87 \times 10^9/\text{mm}^3$. According to Mie Scattering theory, the corresponding SMFPL is 3.68 and 15 μm, respectively, for wavelengths 400 and 800 nm. The optical thickness n is defined as the sample thickness d divided by the SMFPL. For example, if a sample is embedded at a depth of 30 μm, the corresponding optical thickness is 8.14 for wavelength 400 nm. In order to demonstrate the effect of particle size, the SMFPL in a turbid medium consisting of 0.202 μm particles is also assumed to be 3.68 and 15 μm, respectively, for wavelengths 400 and 800 nm. The parameters associated with the two turbid media are summarized in Table 8.1.

The EPSF for 2p fluorescence imaging is shown in Fig. 8.3. It is seen that a peak is shown near the focal center. With increasing the focal depth, the peak becomes less significant and may eventually disappear if the scattering medium is thick enough. The peak is contributed by the fluorescence light excited by ballistic photons that experience no scattering event on their way to the focal plane. The total number of these ballistic photons may be insignificant compared with scattered photons; however, their contribution to the intensity near focal region may still be significant, since these ballistic photons are only distributed near the focal region. Their contribution to the intensity is enhanced due to the quadratic intensity dependence under 2p excitation. If a photon experiences some scattering events, it deviates from the path of ballistic photon and forms a broad distribution on the focal plane.

The EPSF at a depth of 120 μm in a turbid medium consisting of 0.202 μm particles is also illustrated in Fig. 8.3. The EPSF has a stronger central peak, but is more widely distributed in this case, compared with the EPSF at the same depth in a turbid medium consisting of 0.48 μm particles. It is understood that the anisotropy value g is less for smaller scattering particles, which results in a broader distribution of the scattered photons in the EPSF. As a result of this property, the intensity outside the focal region is relatively low in a turbid medium consisting of small scattering particles. Therefore, the central peak becomes more significant due to the quadratic intensity response under 2p excitation.

Table 8.1 Parameters associated with scattering media

ρ (μm)	$\lambda = 400$ nm		$\lambda = 800$ nm	
	g	SMFPL (μm)	g	SMFPL (μm)
0.48	0.89	3.68	0.73	15
0.202	0.69	3.68	0.2	15

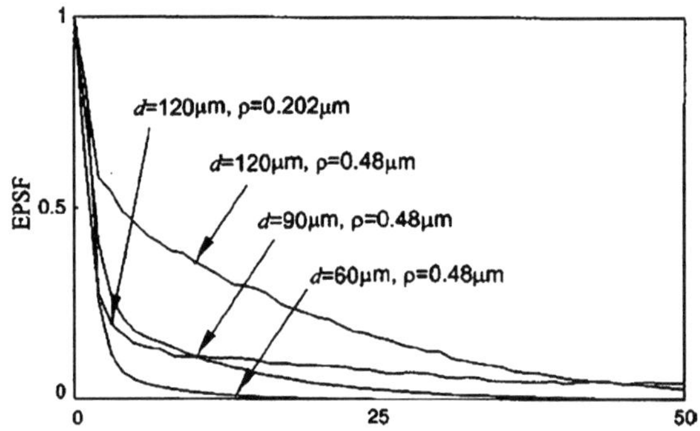

Fig. 8.3 EPSF for 2p fluorescence imaging at different depths *d* for scatterers of diameters 0.48 and 0.202 mm, respectively, ($v_d \to \infty$, NA = 0.25). Reprinted with permission from [10], 2000, American Institution of Physics

8.1.2 Image Resolution

The transverse resolution of an edge object embedded in a turbid medium consisting of either 0.48 μm particles or 0.202 μm particles is illustrated in Fig. 8.4. It is demonstrated in Fig. 8.4 that 2p fluorescence imaging is vastly superior to 1p fluorescence imaging in terms of image resolution if the depth is less than 60 μm. For 2p fluorescence imaging, the transverse resolution stays nearly at the diffraction-limited resolution when $d < 60$ μm, and then quickly deteriorates. The deterioration rate eventually drops approximately after $d = 90$ μm. When $d < 60$ μm, the fluorescence light excited by ballistic light is dominant over that excited by scattered light, and therefore a near-diffraction-limited image resolution can be achieved. In the range of 60 μm $< d <$ 90 μm, the weighting of the fluorescence light excited by ballistic photons drops quickly and the fluorescence light excited by scattered photons becomes strong, thus leading to poor transverse resolution. In this region, the deterioration rate of transverse resolution depends on the decaying rate of ballistic photons. When $d > 90$ μm, the scattered component in the EPSF is dominant in building an image. In this situation, the transverse resolution is decided by the broadness of the EPSF contributed by scattered photons.

A comparison of the two image resolution curves in Fig. 8.4 shows that the transverse resolution in a turbid medium consisting of 0.48 μm particles is better only when $d > 75$ μm, in which case scattered photons dominate the imaging process. In the region $d < 75$ μm, where the central peak in the EPSF dominates the image formation, the resolution is almost the same for the two turbid media.

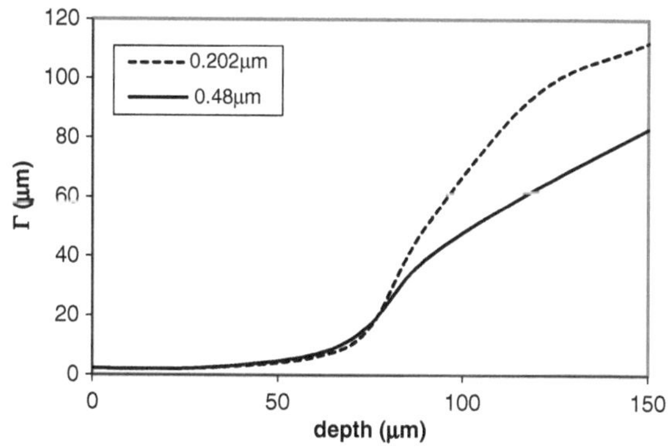

Fig. 8.4 Transverse resolution of a thin edge image as a function of the focal depth for different sizes of scattering particles in 2p fluorescence imaging ($v_d \to \infty$, NA = 0.25). Reprinted with permission from [10], 2000, American Institution of Physics

8.1.3 Signal Level

A comparison of the signal level as a function of the focal depth for 1p and 2p fluorescence imaging is shown in Fig. 8.5. It is shown in Fig. 8.5 that the signal level under 2p excitation drops significantly when the focal depth increases. For example, the signal level drops to only 3 % at d = 30 μm. The signal level for 2p fluorescence imaging decreases faster, compared with the signal level for 1p fluorescence imaging. It is also noticed that the signal level under 2p excitation is more than three orders of magnitude smaller than that under 1p excitation when d = 60 μm. In this situation, the 2p fluorescence signal may be too weak to overcome a noise level from an imaging system. Interestingly, the 2p fluorescence light excited by ballistic photons is still dominant and a near-diffraction-limited resolution can still be obtained at d = 60 μm. Therefore, we may draw the conclusion that 2p fluorescence imaging is mainly determined by ballistic photons, and that the penetration depth is limited by the signal level.

8.1.4 Penetration Depth

To demonstrate the limiting factor on the penetration depth under 2p and 1p excitation, we consider a turbid medium consisting of scattering particles (diameter 0.202 μm) suspended in water. The turbid medium was placed in a glass cell with lateral dimensions of 2 cm × 1 cm. The thickness of the glass cell, d, was varied from 25 up to 250 μm.

A uniform fluorescent polymer bar was embedded at the bottom wall of the glass cell. The fluorescent bar can be excited under 1p (λ_s = 488 nm) and 2p (λ_t = 800 nm) excitation to give a peak fluorescence wavelength of approximately 520 nm. Due to

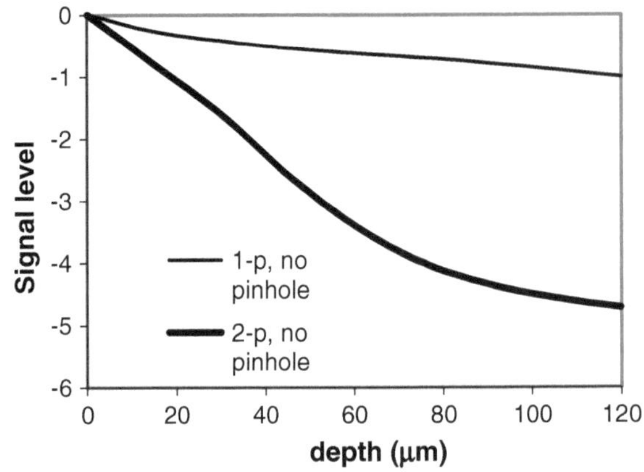

Fig. 8.5 Signal level as a function of the focal depth for 1p and 2p fluorescence imaging, (NA = 0.25, ρ = 0.48 μm). Reprinted with permission from [10], 2000, American Institution of Physics

the different wavelengths associated with 1p excitation, 2p excitation, and fluorescence the scattering cross-section σ_s, the SMFPL l_s and the anisotropy value g are different and can be calculated using Mie scattering theory [14] (Table 8.2).

The prepared sample cell was placed under an Olympus confocal scanning microscope: Fluoview. For 1p excitation, an Ar ion laser at wavelength 488 nm was used. A Spectra-Physics ultrashort pulsed laser: Tsunami, which had a pulse width of 80 fs, was employed for 2p excitation at wavelength 800 nm. To avoid the effect of the refractive-index mismatching between the turbid medium and the cover glass of the cell, a water-immersion objective (Olympus UplanApo 60×, ∞/1.13–0.21, NA = 1.2, working distance = 250 μm) was used. In the case of 1p excitation, a pinhole of 300 μm in diameter was placed in front of the detector to produce an optical sectioning effect with strength similar to that under 2p excitation without using pinhole. The dependence of the signal level and the resolution α on the cell thickness d is depicted in Figs. 8.6 and 8.7 which also includes the Monte Carlo simulation results corresponding to the experimental condition. A good agreement between experiments and theoretical predictions is observed. For the turbid medium we used, the image resolution under 2p excitation is two orders of magnitude higher than that under 1p excitation, while the signal strength attenuates much faster with increasing depth.

Although the above results were obtained for a given size of scatterers, the conclusion regarding the effect of Mie scatterers on 2p excitation holds for other sized Mie scatterers. In particular, the smaller the scatterer size, the smaller the anisotropy value and the larger the SMFPL. Based on this property, we can conclude that for a real tissue medium consisting of different size of Mie scatterers, the 2p fluorescence signal level is lower than 1p signal level at a deep depth. This conclusion is demonstrated in Figs. 8.8 and 8.9. The penetration depth of 2p excitation in imaging through a tissue medium is smaller than that of 1p excitation because of the lower signal level in the former case. However, within the depth of detectable signal, 2p excitation leads to image resolution significantly higher than that under 1p excitation.

Table 8.2 Calculated values of the scattering cross-section σ_s, the scattering mean free path length l_s, and the anisotropy value g for polystyrene beads of diameter 0.202 μm at wavelengths 488, 520, and 800 nm, respectively, for a given particle weight concentration of 2.5 %

Wavelength (nm)	Relative particle size, A (a/λ)	Scattering efficiency, $Q_s = \sigma_s/\sigma_g$	Scattering cross-section, σ_s (μm^2)	SMFPL, l_s (μm)	Anisotropy value, g
488	0.2069	0.1586	5.07×10^{-3}	35.7	0.54
520	0.1942	0.1356	4.34×10^{-3}	41.8	0.482
800	0.1263	0.0389	1.24×10^{-3}	145.38	0.20

The geometric cross-section of the scatterers is 0.032 μm^2

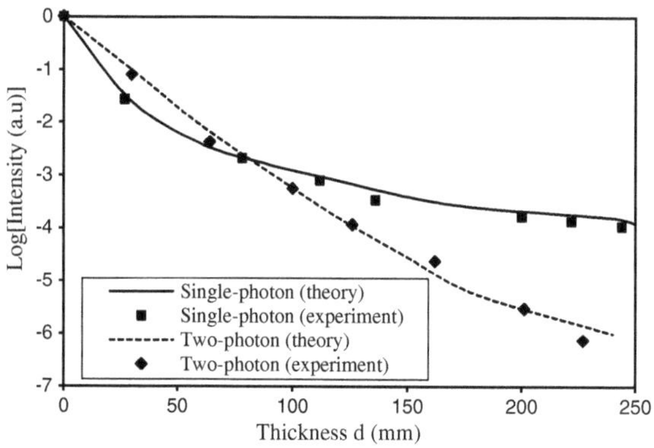

Fig. 8.6 Signal level under 2p and 1p excitation as a function of the penetration depth of a turbid medium consisting of scatterers of diameter 0.202 μm. Reprinted with permission from [11], 2000, American Institution of Physics

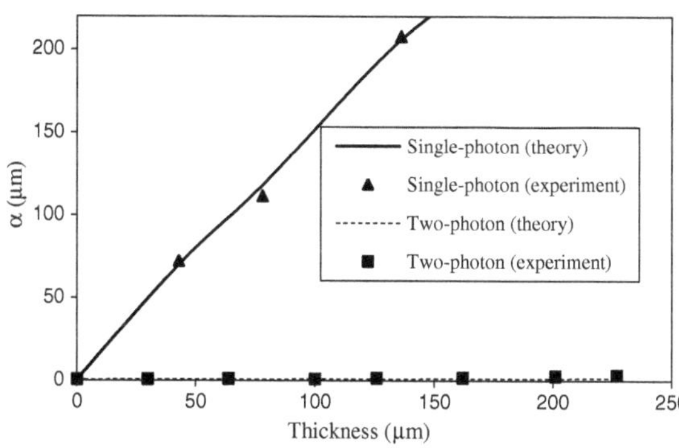

Fig. 8.7 Image resolution under 2p and 1p excitation as a function of the penetration depth of a turbid medium consisting of scatterers of diameter 0.202 μm. Reprinted with permission from [11], 2000, American Institution of Physics

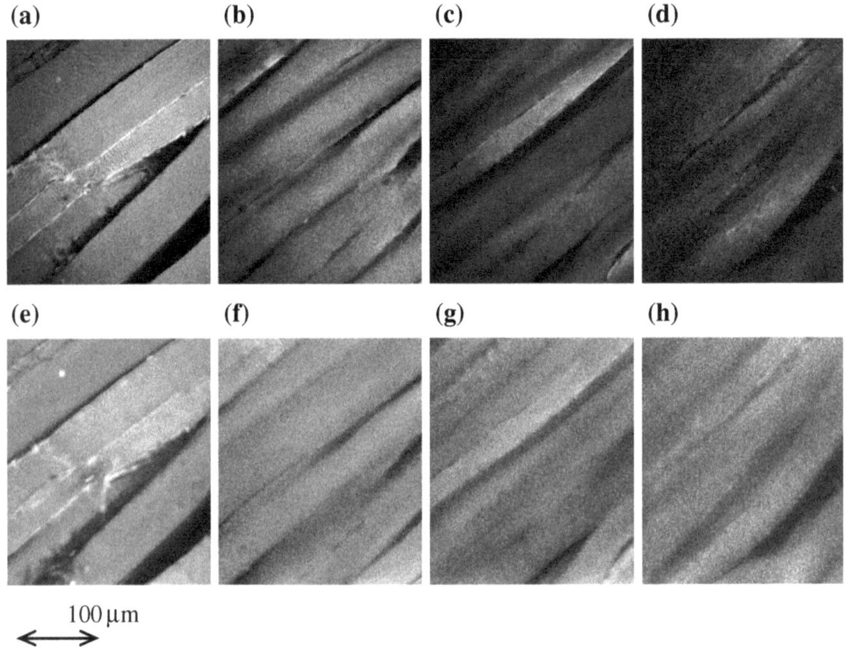

Fig. 8.8 Measured transverse (x–y) images through a thick muscle tissue sample under 2p and 1p excitation. **a–d** 2p fluorescence images at depths of 10, 30, 50, 70 μm; **e–h** 1p fluorescence images at depths of 10, 30, 50, 70 μm. Reprinted with permission from [11], 2000, American Institution of Physics

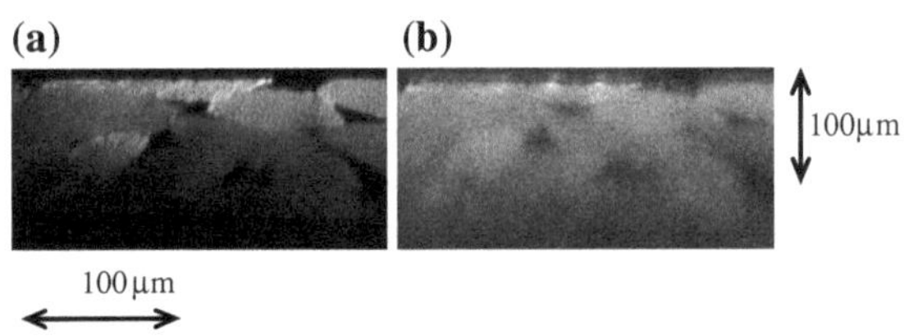

Fig. 8.9 Measured axial (x–z) images through a thick muscle tissue sample under 2p (**a**) and 1p (**b**) excitation. Reprinted with permission from [11], 2000, American Institution of Physics

8.2 Influence of System Parameters

Commonly, the performance of an imaging system is governed by the point spread function which is determined by system parameters such as the NA of the objective lens, the size of a confocal pinhole, etc. In this section, the influence of various parameters associated with a microscopic imaging system under the 2p excitation is discussed in detail.

8.2.1 Numerical Aperture

For the same scattering media described in Sects. 8.1.1–8.1.3, the EPSF at a depth of 120 μm for 2p fluorescence imaging is shown in Fig. 8.10 for different values of the NA of an objective. The EPSF for 2p excitation is not necessarily poorer if a higher NA objective is used. In fact, in this situation, the central peak is more pronounced, but the EPSF appears to be broader.

In Fig. 8.11, the transverse resolution as a function of the focal depth is illustrated for different values of the NA of an objective. It is noted that for $d < 60$ μm, the image resolution is slightly better if a higher Na objective is used. In this region, the central peak in the EPSF is dominant in forming an image. The central peak in the EPSF is mainly contributed by fluorescence excited by ballistic photons. Therefore, a near-diffraction-limited resolution can be achieved in this situation. Since a higher NA objective produces a narrower diffraction spot, the corresponding transverse resolution is better. In the region where $d > 90$ μm, the transverse resolution is better when a lower NA objective is used. In this region, the broad distribution in the EPSF, contributed by scattered photons, is dominant in image formation. In this situation, a high NA objective collects more scattered photons, which results in a broader distribution in the EPSF, and therefore leads to poorer resolution. It is in the region of 60 μm $< d <$ 90 μm where the central peak in the EPSF starts to lose its dominance that a quicker degradation in image resolution occurs for a higher NA objective.

In conclusion, when the ballistic peak in the EPSF is dominant in forming an image, a higher NA objective offers a better resolution. When the broad distribution contributed by scattered photons is dominant, a lower NA objective gives better transverse resolution.

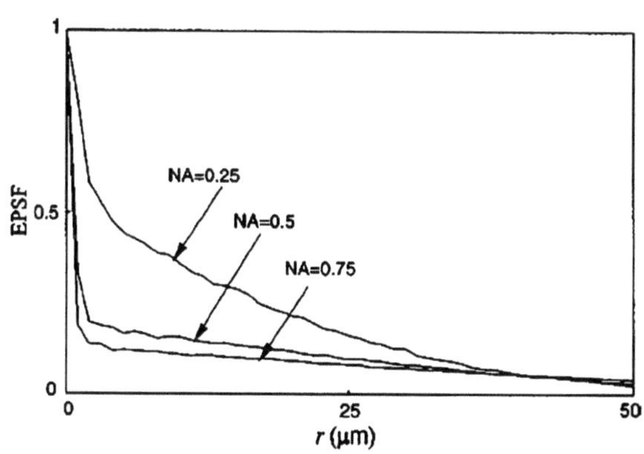

Fig. 8.10 EPSF for different values of the numerical aperture of an objective in 2p fluorescence imaging ($v_d \to \infty$, $\rho = 0.48$ μm). Reprinted with permission from [10], 2000, American Institution of Physics

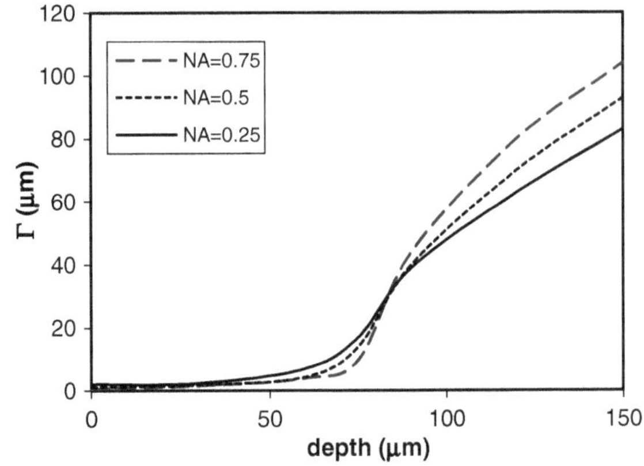

Fig. 8.11 Transverse resolution for different values of the NA of an objective in 2p fluorescence imaging ($v_d \to \infty$, $\rho = 0.48$ μm). Reprinted with permission from [10], 2000, American Institution of Physics

8.2.2 *Confocal Pinhole*

Now we consider the effect of a pinhole on 2p fluorescence imaging. It has been demonstrated in Chap. 7 that a confocal pinhole is efficient in suppressing highly scattered photons, and as a result, it provides a significant improvement in the 1p fluorescence imaging. The EPSF for a pinhole of diameter 100 μm at a depth of 120 μm is illustrated in Fig. 8.12a. It is noted that the utilization of a finite sized pinhole makes little difference in the central peak of the EPSF. However, it does reduce the contribution of scattered photons in the region outside the focal center. This feature implies that a pinhole may not be efficient in improving image quality in 2p fluorescence imaging through turbid media if the central peak is dominant in the EPSF. In Fig. 8.12b, the transverse resolution as a function of the focal depth is illustrated. As expected, the resolution improvement by utilization of a finite sized pinhole is not significant until a turbid medium becomes thicker, $d > 75$ μm, in which case the fluorescence excited by scattered photons is dominant.

The utilization of a pinhole offers little improvement in 2p fluorescence imaging in terms of image resolution (Fig. 8.12b). Moreover, using a pinhole significantly reduces signal strength (Fig. 8.12c), which can be a serious problem in 2p fluorescence imaging through turbid media. For example, at d = 60 μm, the improvement in transverse resolution is only 9.5 % when a pinhole of diameter 100 μm is used. However, the signal strength drops to only 31 % of the signal level when no pinhole is used. It appears that using a pinhole in 2p fluorescence imaging through turbid media is unwise.

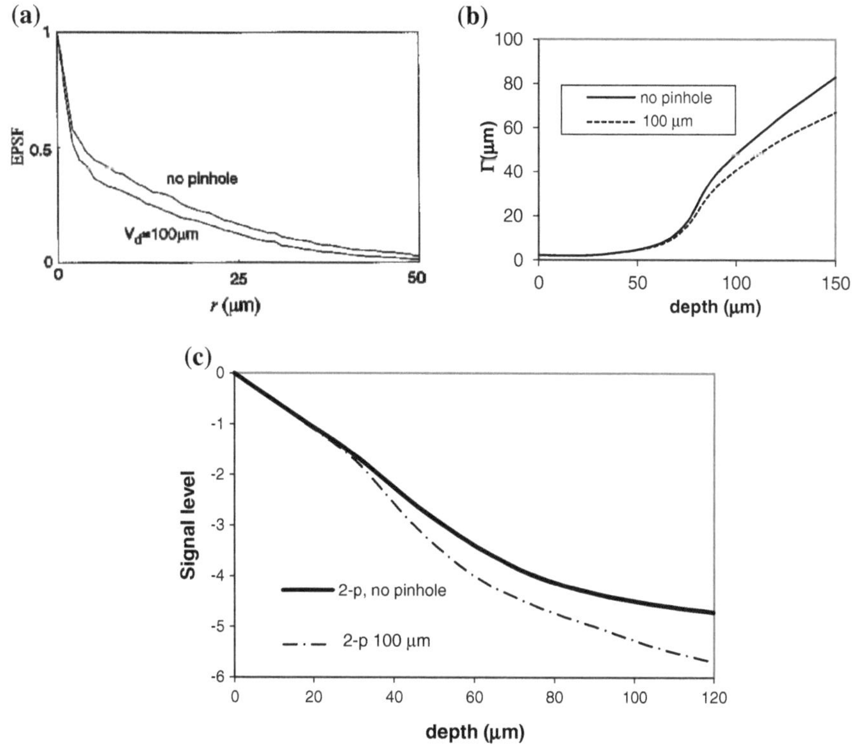

Fig. 8.12 **a** EPSF foa a pinhole of diameter 100 μm at a depth of 120 μm; **b** resolution; **c** signal level for a confocal microscope. Reprinted with permission from [10], 2000, American Institution of Physics

8.3 Two-Photon Imaging Through Complex Scattering Medium Structure

In this section, a detailed investigation into the image resolution and signal level under 2p excitation in scattering media with multiple layer structure is presented.

8.3.1 Multiple Sizes

In Sect. 3.7, the effect of multi-sized scatters has been included in the Monte Carlo method. In this section, a modified Monte Carlo simulation model considering inhomogeneous turbid media with scatterers of multiple sizes is established for studying multi-photon fluorescence microscopy [10–13].

A real biological tissue medium may be complicated [14–18]. Here a single-layer inhomogeneous turbid medium consisting of n types of spherical scatterers is considered. Each type of scatterer has a given size (diameter) ρ_i and concentration c_i. A SMFPL (l_i) for each type of scatterer is determined by its concentration (c_i) and its corresponding scattering cross-section (σ_{si}), based on Mie theory:

$$l_i = 1/(c_i\sigma_{si}) \tag{8.3}$$

To determine the scattering step length between each two consecutive scattering events of a photon in a turbid medium with multi-size scattering particles, we first independently calculate the scattering step length (s_i) from each type of particles according to the following equation:

$$s_i = -l_i \ln(\xi) \quad (i = 1, 2, \ldots, n) \tag{8.4}$$

where $0 < \xi < 1$ is a randomly produced uniform distribution number. Then the shortest step length (s_m) is taken to be the length within which a photon propagates freely. The anisotropy value (g_m) corresponding to the m-th type of particles, which is also calculated based on Mie theory [19], is used to determine the scattering direction of the photon. The accuracy of this method has been verified by our previous model when there exists only one type of particles [10, 12].

For understanding image performance of a turbid medium of multi-size scattering particles, a parameter, the effective mean free path length (l'), is introduced as a measure of the randomness in such an inhomogeneous medium

$$1/l' = \sum_{i=1}^{n} 1/l_i \tag{8.5}$$

The effective mean free path length (l') weights the contributions of scattering cross-sections from different types of particles to the scattering features of the turbid medium. It is an analogous parameter to the mean free path length (l) in a homogeneous medium which has scattering particles of one size.

According to the size distribution of small organelles and nuclei, three groups of inhomogeneous turbid media are considered for understanding image formation. Groups S and L, which attempt to investigate the scattering features from small organelles and large nuclei in cells, contain nine types of scattering particles. Their diameter (ρ) covers from 0.1 to 0.9 μm and from 2.6 to 3.4 μm, respectively, with a diameter interval of 0.1 μm and the mean size of these two groups is 0.5 and 3.0 μm, respectively. The third group of the turbid media (M) is a mixture of groups S and L to simulate the situation where small scattering organelles and large nuclei both exist. Based on Mie theory, the corresponding anisotropy value (g_i) and scattering cross-section (σ_{si}) of each type of particles under 2p excitation are shown in Table 8.3.

The NA of the objective is chosen to be 0.25. It is assumed that the wavelength of the excitation beam, λ_{ex}, under 2p excitation is 800 nm and the fluorescence wavelength λ_{fluo} is 400 nm.

Table 8.3 Anisotropy values (g_i) and scattering cross-sections (σ_{si}) of different types of particles under 2p excitation

l' (μm)		0.1	0.2	0.3	0.4	0.5	0.6	0.7	0.8	0.9
400 nm (1p excitation) (Fluorescence)	σ_{si} (μm^2)	0.000296	0.0081	0.0483	0.155	0.36	0.686	1.132	1.678	2.28
	g_i	0.193	0.685	0.804	0.864	0.894	0.916	0.924	0.928	0.929
800 nm (2p excitation)	σ_{si} (μm^2)	0.000023	0.00128	0.00877	0.0324	0.0907	0.193	0.366	0.619	0.971
	g_i	0.047	0.193	0.450	0.685	0.739	0.804	0.844	0.864	0.885
l' (μm)		2.6	2.7	2.8	2.9	3.0	3.1	3.2	3.3	3.4
400 nm (1p excitation) (Fluorescence)	σ_{si} (μm^2)	14.28	15.1	15.65	16.16	16.62	16.56	16.23	15.98	16.44
	g_i	0.889	0.896	0.892	0.877	0.863	0.863	0.870	0.877	0.872
800 nm (2p excitation)	σ_{si} (μm^2)	16.45	16.6	16.88	16.91	16.79	16.92	16.56	16.85	16.71
	g_i	0.906	0.900	0.892	0.883	0.874	0.863	0.852	0.844	0.829

For a comprehensive understanding of the effect of the particle concentration c_i, two types of the concentration distributions, the uniform and Gaussian distribution, are investigated, represented by U or G. All concentration distributions discussed in this paper are centrally symmetrical through the mean size particle. Each turbid medium is labelled by two letters followed by two digitals. The first letter represents the type of particle group, while the second letter denotes the type of the concentration distribution. The two digitals correspond to a width parameter δ. The width parameter δ in the medium with a uniform distribution is $\delta_p/2(\delta_p = \rho_{max} - \rho_{min})$ where ρ_{max} and ρ_{min} are the maximum and minimum particle sizes, respectively. In the medium with a Gaussian distribution, the width parameter δ is defined as the full width at half maximum of the concentration distribution. For example, the SG01 medium contains a small group of particles with a Gaussian distribution and $\delta = 0.1$ μm.

In order to observe scattering features of different concentrations, it is assumed that the averaged total geometrical cross-section (σ) is constant in all media, which can be mathematically expressed as

$$\int \pi(\rho/2)^2 c(\rho)\,\mathrm{d}\rho = \pi(\rho_0/2)^2 c_0 \tag{8.6}$$

where ρ is the diameter for a particle type existing in an inhomogeneous medium and $c(\rho)$ is the corresponding concentration. Here ρ_0 refers to the scattering particle size in a homogeneous medium with concentration c_0. In our calculation, $\rho_0 = 0.5$ μm and $c_0 = 0.6/\mu m^3$ are selected for the uniform S group, while for the uniform L group, $\rho_0 = 3$ μm and $c_0 = 0.0167/\mu m^3$ are chosen.

According to the concentration distributions and the scattering cross-section at different excitation wavelengths (Table 8.5), the effective mean free path length (l') of the various inhomogeneous media under 1p, 2p, and three-photon (3p) excitation can be calculated by (8.3) and (8.5) and are summarized in Table 8.4.

The EPSF represents the performance of an imaging system including a turbid medium. As an example, EPSFs for turbid media SU00, SU04, and SG04 at different focal depths under two-photon (2p) excitation are demonstrated in Fig. 8.13. The sharp peak shown in Fig. 8.13 demonstrates the existence of ballistic and less scattered snake photons. It is noticed that as the focal depth increases, the EPSF becomes broad and the peak value decreases. This is because as the focal depth increases, the scattering events increase, which leads to the broadening of the EPSF. Meanwhile, due to strong scattering, the number of ballistic and snake photons that undergo few scattering events around the center of the focus region decreases dramatically. As a result, the peak value reduces.

The transverse image resolution and signal level in a series of SU media under 2p excitation are first demonstrated in Fig. 8.14. Although the total geometrical cross-sections of the size distributed turbid media are equal, optical microscopic image formation in those media is affected differently due to the different scattering features of the scatterers of various sizes. The more types of particles around the mean size of 0.5 μm are included, the worse the image performance on resolution

Table 8.4 The effective mean free path length (l') of the various inhomogeneous media under 1p, 2p and 3p excitation

Type of medium		l' (μm)		Type of medium		l' (μm)	
		400 nm (1p excitation) (Fluorescence)	800 nm (2p excitation)			400 nm (1p excitation) (Fluorescence)	800 nm (2p excitation)
SU	SU00/SG00	4.6296	18.3756	LU	LU00/LG00	3.6029	3.5664
	SU01	4.2699	16.2234		LU01	3.6850	3.5918
	SU02	3.7832	13.0401		LU02	3.6863	3.5618
	SU03	3.3224	10.3066		LU03	3.7573	3.5907
	SU04	2.9837	8.2988		LU04	3.8000	3.6071
SG	SG01	4.5736	18.0067	LG	LG01	3.5956	3.5501
	SG02	4.3182	16.2384		LG02	3.6360	3.5524
	SG03	3.9779	13.9597		LG03	3.7065	3.5959
	SG04	3.6824	12.0424		LG04	3.6324	3.5029
M	SG04 + LU02	1.8422	2.7488				

Table 8.5 Anisotropy value g and scattering mean free path length l_s for 1p and 2p excitation and fluorescence wavelengths

ρ (μm)	1p excitation		2p excitation		Fluorescence	
	$\lambda_{ex} = 400$ nm		$\lambda_{ex} = 800$ nm		$\lambda_{fluo} = 400$ nm	
	g	l_s (μm)	g	l_s (μm)	g	l_s (μm)
0.48 (L)	0.89	3.68	0.73	15	0.89	3.68
0.202 (S)	0.69	3.68	0.2	15	0.69	3.68

and signal level. This situation may be explained by the effective mean free path length (l') introduced in this paper. Table 2 clearly shows that in the medium which has the δ value of 0.4 μm (SU04), l' has the shortest value under 2p excitation. Figure 8.15 shows the transverse resolution and signal level in a series of the LU media under 2p excitation. Unlike the situation in the SU media, the difference of the transverse resolution and signal level between different types of the media is not significant under three types of excitation. Figure 8.16 shows the transverse resolution and the signal level in the SG media. Compared with those in the SU media (Fig. 8.14), the transverse resolution and signal level under 2p excitation are better in the SG media. An explanation can be obtained from the fact that l' in the SG media is larger than that in the SU media under multi-photon excitation. This feature is caused by the suppression of the effect of the large particles in the case of Gaussian distribution. This feature can also be reflected from the EPSFs shown in Fig. 8.13. At the same focal depth, the EPSF in SG04 is narrower than that in SU04 under 2p excitations. Similarly, for the L group media of a Gaussian concentration distribution, the transverse resolution and signal level are shown in Fig. 8.17. There is little difference in transverse resolution and signal level between the LG and LU media under 2p excitation.

A mixture medium is investigated to simulate the combined effect of uniform and Gaussian distributions. In this simulation, it is supposed that there is a diverse distribution of particles of small sizes and a relatively concentrated distribution of

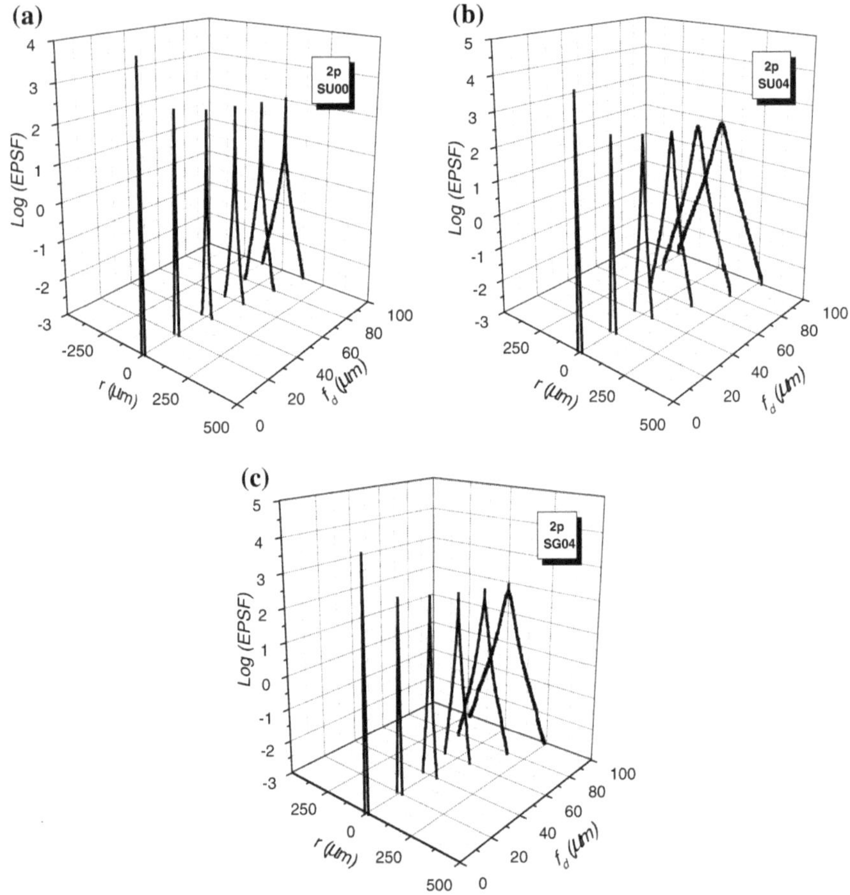

Fig. 8.13 EPSFs for fluorescence imaging at different focal depths in SG00, SG04, and SU04 turbid media under 2p excitation. Reprinted with permission from [25], 2003, SPIE

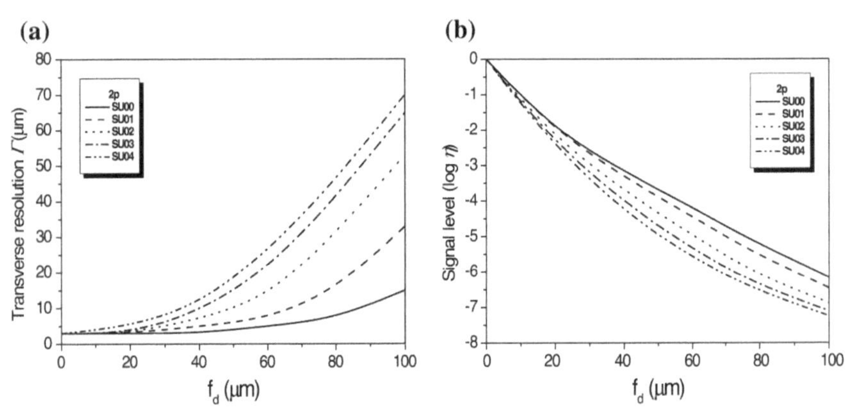

Fig. 8.14 Transverse image resolution and signal level as a function of the focal depth in the SU media under 2p excitation. Reprinted with permission from [25], 2003, SPIE

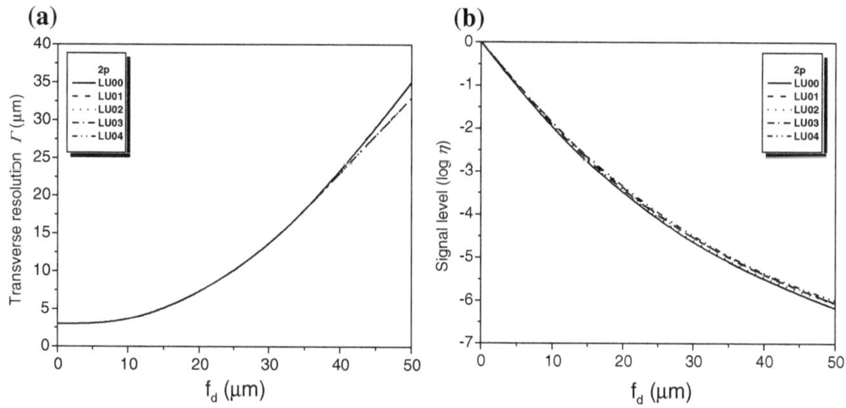

Fig. 8.15 Transverse image resolution and signal level as a function of the focal depth in the SU media under 2p excitation. Reprinted with permission from [25], 2003, SPIE

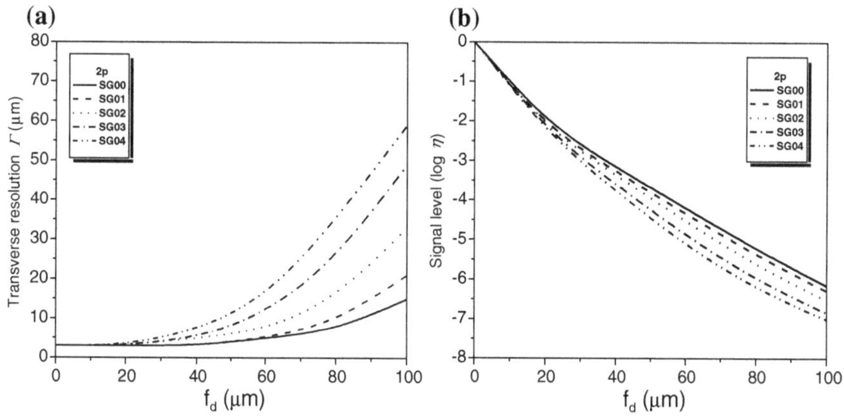

Fig. 8.16 Transverse image resolution and signal level as a function of the focal depth in the LG media under 2p excitation. Reprinted with permission from [25], 2003, SPIE

particles of large sizes. This condition corresponds to the situation in cells where small scatterers (such as small organelles) distribute diversely, while the size of the scattering particles like nuclei has relatively small variation. It is noticed from Figs. 8.15 and 8.17 that for the L group, the imaging performance between the uniform and Gaussian distribution is not significant and the effect of the size distribution is not pronounced. Therefore, in the mixture medium M, we arbitrarily choose LU02 to be a simulation medium of nuclei. On the other hand, SG04 is selected to simulate the maximum effect of the size distribution of small scatterers. However, in order to keep the equality of the total geometrical cross, this mixture medium M has half the concentration values for each of its corresponding

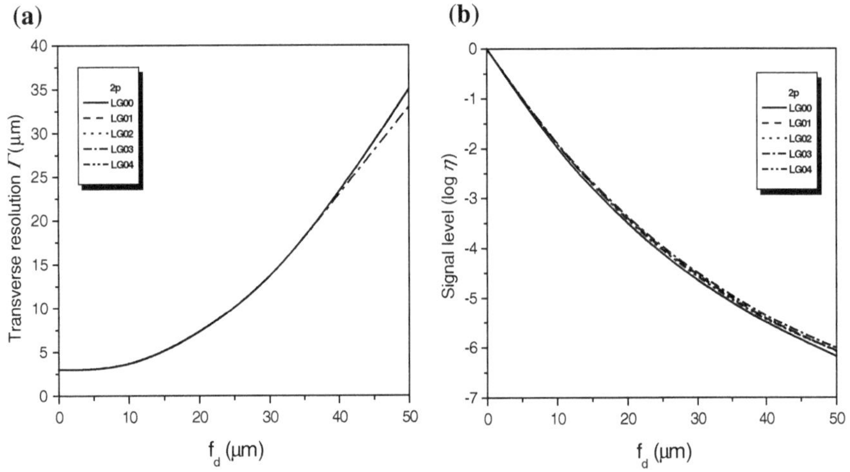

Fig. 8.17 Transverse image resolution and signal level as a function of the focal depth in the LU media under 2p excitation. Reprinted with permission from [25], 2003, SPIE

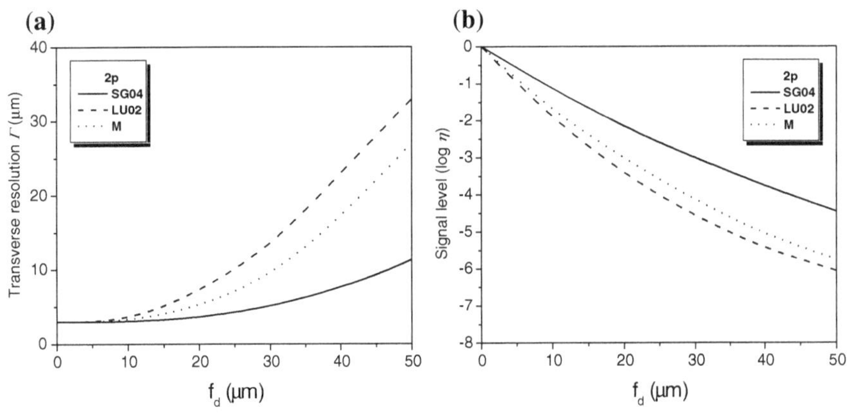

Fig. 8.18 Transverse image resolution and signal level as a function of the focal depth in the M media under 2p excitation. Reprinted with permission from [25], 2003, SPIE

constitution scatterers compared to those in the media SG04 and LU02. Figure 8.18 demonstrates the image transverse resolution and signal level in this mixture medium M. For comparison, transverse resolution and signal level in the SG04 and LU02 media are also demonstrated in the same plots. It is noticed that image performance in a mixture medium M (either transverse resolution or signal level) falls in between two extreme cases in which turbid medium only consists of either small or large particles and the image performance is more affected by the LU02 medium than the SG04 medium under three types of excitation.

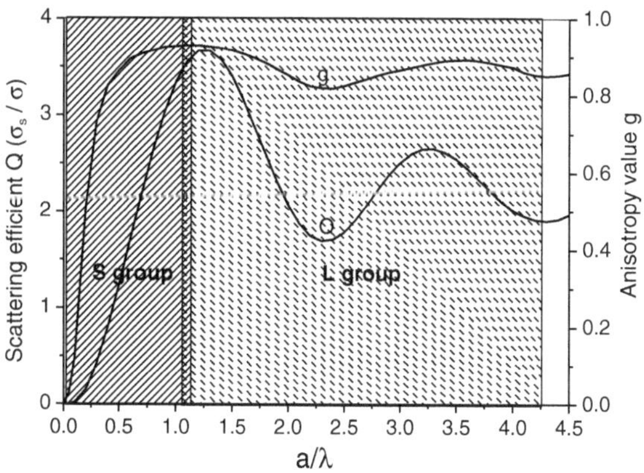

Fig. 8.19 Scattering efficiency Q and anisotropy value g as a function of the scattering parameter a/λ. The refractive indices of scatterers (spherical particles) and the immersion medium (water) are 1.59 and 1.33, respectively (Reprinted with permission from [25], 2003, SPIE)

These phenomena can be explained from the relationship of the scattering efficiency Q and the anisotropy g value to a scattering parameter (a/λ) (Fig. 8.19). The scattering efficiency Q is defined as the ratio of the scattering cross-section (σ_s) to the geometrical cross-section (σ) and a is the radius size of a scattering particle. The smaller scattering efficiency implies a reduction of multiple scattering events, which leads to high image resolution and signal level. In our case, the S media include particles that have relative sizes, a/λ, within the ranges from 0.125 to 1.125, 0.0625 to 0.5625 for the illumination and excitation wavelengths, respectively. While the L media include the relative particles sizes within the ranges from 3.25 to 4.25 and 1.625 to 2.125 the illumination and excitation wavelengths. Thus the scattering Q changes significantly as a/λ falls within the range of 0.0417–1.125 in the S media and shows a slight variation while a/λ falls within the range of 1.083–4.25 in the L media, as indicated in Fig. 8.19. This feature provides an understanding of the significant influence of the scattering features in the S media. In the S media, the Q value of the largest particles (0.9 μm) is about 500 times that of the smallest particles (0.1 μm) under 2p excitation. This effect results in a significant impact on the image performance.

8.3.2 Multiple-Layer Structures

So far, the turbid media considered in the current microscopic Monte Carlo simulation are a single-layer structure. However, biological tissue usually exhibits a complex layer structure such as skin tissue. Although 2p fluorescence microscopy has been used to image through skin tissue, the penetration depth and the limit of image resolution under multi-photon excitation cannot be determined from the

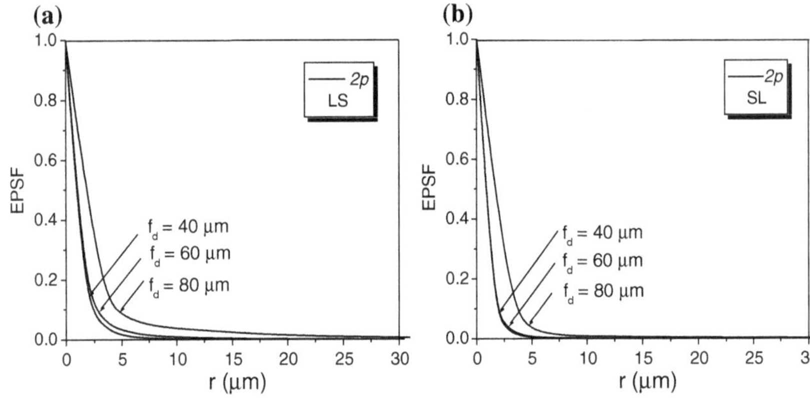

23

Fig. 8.20 EPSF for 2p fluorescence imaging at different focal depths in the double-layer turbid medium: **a** LS medium; **b** SL medium. Reprinted with permission from [12], 2002, American Institution of Physics

current Monte Carlo simulation model. In this section, a detailed investigation into the image resolution and signal level under 2p excitation in scattering media with multiple layer structure is presented.

To compare the results of a double-layer turbid medium with those from a single-layer turbid medium, we choose two kinds of spherical scattering particles suspended in water. The first kind, called large particles (L), has a diameter of 0.48 μm, and the second, called small particles (S), has a diameter of 0.202 μm. Each layer in the double-layer structure consists of either large particles or small particles. If the top layer (L_1) includes large particles and the bottom layer (L_2) includes small particles, the double-layer turbid structure is called the LS medium. Otherwise, it is called the SL medium. The thickness of the two layers of the media is assumed to be 60 μm. According to Mie Scattering theory [19], the corresponding anisotropy value g and SMFPL l_s are shown in Table 8.3.

Figure 8.20 gives the EPSFs for 2p fluorescence imaging at the three focal depths in the LS and SL turbid media. It can be seen that the EPSF in both LS and SL media does not change appreciably and that its width is almost close to that given by the diffraction theory when the focal depth is less than 60 μm. This feature is caused by the fact that the contribution from ballistic photons to 2p fluorescence emission may be dominant in the focal region at the depth 4–5 times of the SMFPL. Figure 8.21 shows the comparison of the EPSFs between LS and SL media at a given focal depth. It is seen that unlike the situation under 1p excitation, the EPSF in the SL medium under 2p excitation is slightly narrower than that in the LS medium. This property is consistent with the previous result in the single-layer turbid medium under 2p excitation [9]. For the given SMFPL (Table 8.3), the small anisotropy value g in the S layer results in a broad distribution of scattered photons and their contribution to intensity is reduced because of the quadratic intensity dependence under 2p excitation. As a result, the central peak mainly resulting from

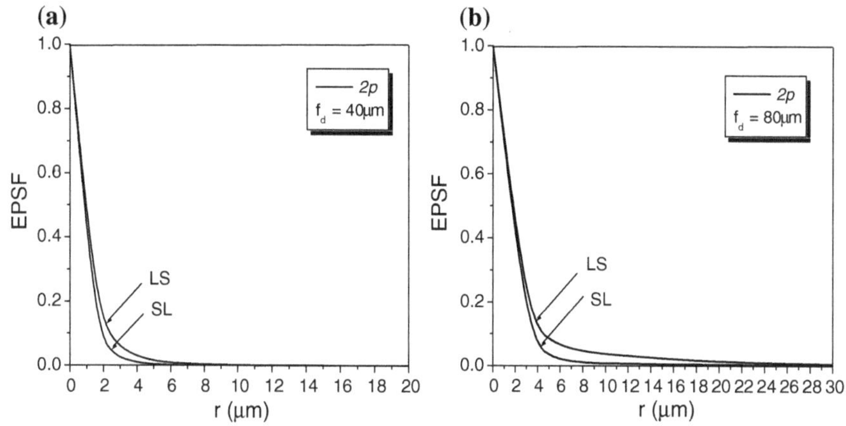

Fig. 8.21 Comparison of EPSFs for 2p fluorescence imaging between LS and SL media. **a** $f_d = 40$ μm; **b** $f_d = 80$ μm. Reprinted with permission from [12], 2002, American Institution of Physics

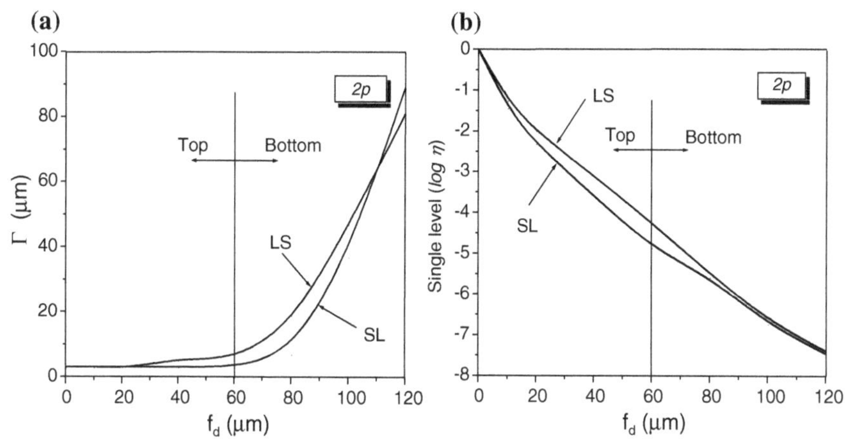

Fig. 8.22 Transverse image resolution (**a**) and signal level (**b**) as a function of the focal depth in the LS and SL media under 2p excitation. Reprinted with permission from [12], 2002, American Institution of Physics

ballistic photons in the SL medium becomes slightly narrower, compared with that in the LS medium.

Transverse image resolution as a function of the focal depth in the LS and SL media under 2p excitation is shown in Fig. 8.22. The transverse resolution in the top layer of the LS and SL media is almost close to the diffraction-limited value, as expected from Fig. 8.19. Below the interface, the resolution is degraded quickly in the both cases, which means that the contribution from scattered photons is significantly increased. To understand the behavior near the interface, we define the rate of resolution degradation, β, as $\beta = \mathrm{d}\Gamma/\mathrm{d}(f_\mathrm{d})$. The rate of resolution degradation β immediately below the interface is caused by two processes. The first one is the change in the anisotropy value g. In general, photons in the S layer (smaller anisotropy value g) are statistically scattered into a larger angle than those in the L layer (larger anisotropy value g) [12]. It has been demonstrated that the photons scattered at a larger angle lead to lower resolution than the photons scattered at a smaller angle. As a result, the increase of the anisotropy value g from the S layer to the L layer in the SL medium results in the decrease of β near the interface. The second process is a significant contribution from the scattered photons when the focal depth is approximately 60 μm, which results in the increase of β in both the media. The combination of these two processes gives rise to the fact that the degradation of resolution in the LS medium is faster than that in the SL media when the focal depth moves from the top layer to the bottom layer. It is seen from Fig. 8.22b that the signal level drops faster in the S medium, however, after penetrating the same distance in L and S media, the signal level curves representing the LS and SL structures converges.

8.4 Three-Photon Fluorescence Imaging

In the previous sections, it has been demonstrated that 2p fluorescence imaging holds advantage over the 1p fluorescence imaging in terms of achieving high resolution, because of utilizing longer excitation wavelength and the quadratic dependence of the fluorescence on the excitation intensity. It is expected that high-order nonlinear effects, such as 3p excitation can further enhance these advantages. In this section, 3p fluorescence imaging through highly scattering media is investigated in detail. Based on the Monte Carlo simulation model developed in Sect. 3.9, we use it to calculate the resolution and signal level in human skin tissue under 1p, 2p, and 3p excitation. Human skin tissue is a complex and highly scattering thick tissue. It can be considered to be a double-layer structure mainly consisting of epidermis and dermis [20]. We assume that the wavelength is 365, 730 and 1,095 nm for 1p, 2p, and 3p excitation, respectively, and that the fluorescence wavelength is 450 nm. The absorption and scattering parameters of skin tissue at these wavelengths are summarized in Table 8.6. According to the anatomical

Table 8.6 Absorption and scattering parameters of skin tissue under 1p, 2p, and 3p excitation (Reprinted with permission from [12], 2002, American Institution of Physics)

Skin layers	Epidermis			Dermis		
	μ_a (cm^{-1})	g	l_s (μm)	μ_a (cm^{-1})	g	l_s (μm)
1p Excitation (λ_{ex} = 365 nm)	100	0.72	9.1	7	0.72	21.8
2p Excitation (λ_{ex} = 730 nm)	39	0.83	23.3	2.4	0.83	55.6
3p Excitation (λ_{ex} = 1,095 nm)	0.87 [21]	0.9	84.5 [21]	0.87 [21]	0.9	84.5 [21]
Fluorescence (λ_{fluo} = 450 nm)	58	0.75	14.3	4.1	0.75	35.1

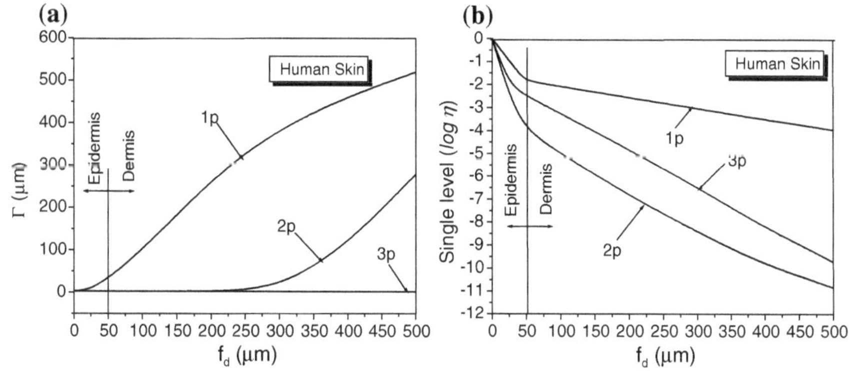

Fig. 8.23 Comparison of transverse image resolution (**a**) and signal level (**b**) in skin tissue under 1p, 2p, and 3p excitation. Reprinted with permission from [12], 2002, American Institution of Physics

structure of the human skin [20], it can be assumed that the thickness of the epidermis and dermis layers is 50 and 450 μm, respectively.

The image resolution Γ and the signal level η of the human skin double layers under three excitation situations are shown in Fig. 8.23a, b, respectively. Because of the big difference in the SMFPL under 1p and 2p and 3p excitation, Γ and η behave differently. Figure 8.23a shows that the resolution under 3p excitation is the best; a diffraction-limited resolution can be kept within the whole thickness of the double-layer skin structure. However, such a diffraction-limited resolution value can be maintained only up to a focal depth of 250 μm (corresponding to a depth of 200 μm in the dermis layer) under 2p excitation. Under 1p excitation, even within the epidermis layer, the resolution is worse than that of the diffraction-limited value. A comparison of the signal level under 1p, 2p and 3p excitation (Fig. 8.23b) shows that the signal level under 3p excitation is between those under 1p and 2p excitation. Within the dermis layer, the 3p signal level is approximately 1–2 orders of magnitude higher than that of 2p excitation. Consequently, Fig. 8.23 suggests that 3p excitation is a better choice for fluorescence imaging through the human skin tissue than 2p excitation.

Another example is the human cortex, which aims at investigating neuron activities. Brain cortex has a double-layer structure that is composed of gray matter and white matter, with gray matter lying on the top. Gray matter contains neurons and provides the actual processing capacity, while white matter provides communications between different gray matter areas and between gray matter and the rest of the tissue. The investigation of the penetration depth through human brain cortex under multi-photon microscopy is important for studying human brain disease in biopsy and autopsy and human brain functions [21–25].

According to a recent MRI study, the thickness of the gray matter layer is between 1,000 and 4,500 μm, and the average thickness is approximately 2,500 μm [26, 27]. In this section, the thickness of the gray and white matter layers is assumed to be

Table 8.7 Absorption and scattering parameters of human cortex under 1p, 2p, and 3p excitation (μ_a absorption coefficient, μ_s scattering coefficient, l scattering mean free path length; g anisotropy value) (Reprinted with permission from [24], 2003, Optical Society of America)

Human cortex	Excitation wavelength			
		400 nm	800 nm	1,200 nm
Gray matter	μ_a (1/mm^{-1})	0.25	0.05	0.017
	μ_s (1/mm^{-1})	12.5	7.8	5.53
	l (μm)	78.5	127.4	181
	g	0.85	0.87	0.91
White matter	μ_a (1/mm^{-1})	0.3	0.17	0.1
	μ_s (1/mm^{-1})	42	38	40.2
	l (μm)	23.6	26.2	25
	g	0.75	0.86	0.89

1,000 μm, respectively, for the thorough investigation of image performance through both gray matter and white matter. This human cortex structure is named Cortex1. The excitation wavelength for 1p, 2p and 3p excitation is assumed to be 400, 800, and 1,200 nm, respectively, and the excited fluorescence wavelength is assumed to be 400 nm in the three cases. Optical parameters of gray matter and white matter in human cortex [23] adopted in this simulation are summarized in Table 8.7.

The 1p, 2p, and 3p fluorescence EPSFs at the focal depth of 800, 1,000, and 1,200 μm in Cortex1 are shown in Fig. 8.23 for an objective of NA 0.25. The focal depth f_d is defined as the distance between the surface of the gray matter layer to the focal plane and r is the radial distance in the focal plane. It is seen that in all the three cases, the EPSF under 1p excitation is the broadest and its difference from those under 2p and 3p excitation is significant, especially within the gray matter layer (Fig. 8.24a) and on the boundary (Fig. 8.24b). The EPSFs under 2p and 3p excitation within the gray matter layer (Fig. 8.24a) are diffraction-limited and the difference between them is undistinguishable. At the boundary, the difference in the EPSF between 2p and 3p excitation gradually becomes large and observable (see the inset of Fig. 8.24b), showing the narrower EPSF under 3p excitation. When the focus moves into the white matter layer, the EPSF under 1p, 2p and 3p excitation dramatically becomes broad especially for those under 2p and 3p excitation (Fig. 8.24c). The reason for this feature is that the white matter has a shorter mean free path length (l) and the smaller anisotropy value (g) compared with those in the gray matter, which results in much stronger scattering and accordingly scattered photons become dominant in image formation.

Image resolution Γ and signal level η under 1p, 2p, and 3p excitation along the entire thickness Cortex1 are shown in Fig. 8.25a, b, respectively. It is understandable from Fig. 8.25a that only within very shallow depth in gray matter ($f_d < 250$ μm), the image resolution under 1p excitation is better than 100 μm. The resolution under 2p and 3p excitation in the gray matter layer keeps almost the near-diffraction limit values and has a slight difference. Compared with the mean free path length under 1p excitation, the mean free path length under 2p and 3p excitation within gray matter is much longer. When the focus moves from the gray matter into the white matter, the resolution becomes degraded rapidly under both 2p

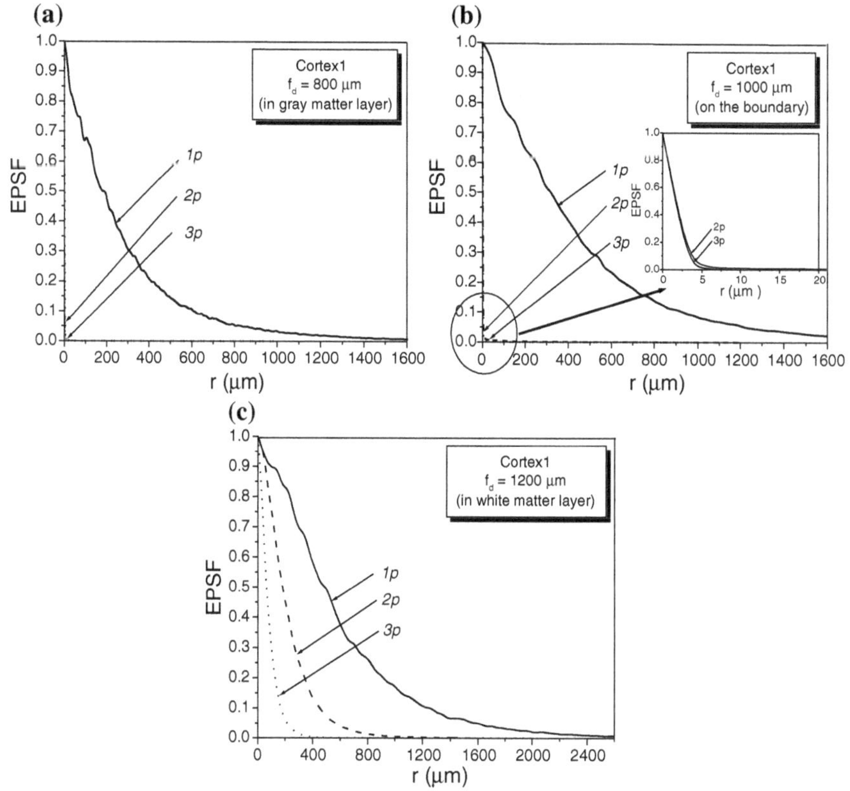

Fig. 8.24 Comparison of 1p, 2p, and 3p fluorescence EPSFs in Cortex1 (NA is 0.25). **a** At the focal depth of 800 μm (within the gray matter layer); **b** at the focal depth of 1,000 μm (on the boundary); **c** at the focal depth of 1,200 μm (within the white matter layer). Reprinted with permission from [24], 2003, Optical Society of America

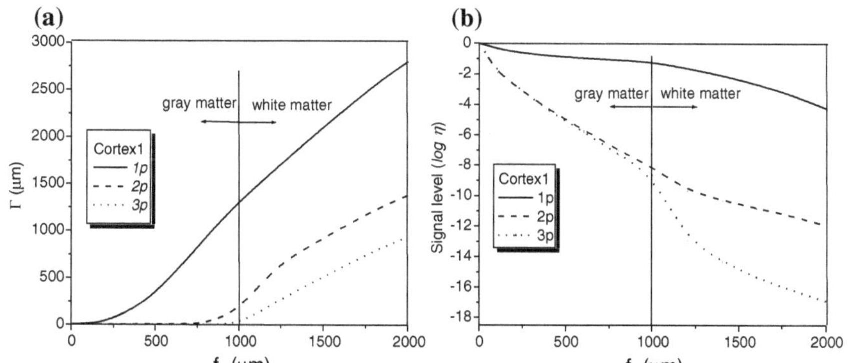

Fig. 8.25 Transverse resolution Γ (**a**) and signal level (**b**) as a function of the focal depth in Cortex1 under 1p, 2p and 3p excitation (NA is 0.25). Reprinted with permission from [24], 2003, Optical Society of America

and 3p excitation due to the significant reduction in the mean free path length from the gray matter to the white matter. 3p excitation results in the best image resolution along the whole range of the sample thickness, showing the fact that Γ in 3p excitation is approximately 1.5 times better than that under 2p excitation. The diffraction-limited resolution under 3p excitation can be maintained almost up to the depth of 1,000 μm (the whole range of the gray matter layer). However, under 2p excitation, the diffraction-limited resolution can be maintained only within a depth of 750 μm.

For signal level in Cortex1 (Fig. 8.25b), it is noticed that signal level under 1p excitation keeps the best situation. The signal level under 2p and 3p excitation has almost the same value until the depth of 1,000 μm (on the boundary), which means that within whole the gray matter region, 2p and 3p excitation makes no difference between the signal levels. However, when the focal depth enters the white matter layer, the signal level under 3p excitation drops much more quickly than that under 2p excitation. This property indicates that the strong scattering due to the short mean free path length and the small anisotropy value in the white matter has a more significant impact on the signal level under 3p excitation than that under 2p excitation.

References

1. W. Denk, J.H. Strickler, W.W. Webb, Two photon laser scanning fluorescence microscopy. Science **248**, 73 (1990)
2. S.P. Schilders, M. Gu, Three-dimensional autofluorescence spectroscopy of rat skeletal muscle tissue under two-photon excitation, Appl. Opt. **38**, 720 (1999)
3. Y. Guo, Q.Z. Wang, N. Zhadin, F. Liu, S. Demos, D. Calistru, A. Tirksliunas, A. Katz, Y. Budansky, P.P. Ho, R.R. Alfano, Two-photon excitation of fluorescence from chicken tissue. Appl. Opt. **36**, 968 (1997)
4. J.M. Schmitt, K. Ben-Letaief, Efficient Monte Carlo simulation of confocal microscopy in biological tissue. J. Opt. Soc. Am. A **13**, 952 (1996)
5. X. Gan, S.P. Schilders, M. Gu, Image formation in turbid media under a microscope. J. Opt. Soc. Am. A **15**, 2052 (1998)
6. X. Gan, S. Schilders, M. Gu, Combination of annular aperture and polarization gating methods for efficient microscopic imaging through a turbid medium: theoretical analysis. Microsc. Microanal. **3**, 495 (1997)
7. C.M. Blanca, C. Saloma, Monte Carlo analysis of two-photon fluorescence imaging through a scattering medium, Appl. Opt. **37**, 8092 (1998)
8. V. Daria, C.M. Blanca, O. Nakamura, S. Kawata, C. Saloma, Image contrast enhancement for two-photon fluorescence microscopy in a turbid medium. Appl. Opt. **37**, 7960 (1998)
9. X. Gan, M. Gu, Spatial distribution of single-photon and two-photon fluorescence light in scattering media: Monte Carlo simulation. Appl. Opt. 39, 1580 (2000)
10. X. Gan, M. Gu, Fluorescence microscopic imaging through tissue-like turbid media. J. Appl. Phys. **87**, 3214 (2000)
11. M. Gu, X. Gan, A. Kisteman, M.G. Xu, Comparison of penetration depth between two-photon excitation and single-photon excitation in imaging through turbid tissue media, Appl. Phys. Lett. **70**, 1551 (2000)
12. X. Deng, X. Gan, M. Gu, Multi-photon fluorescence microscopic imaging through double-layer turbid tissue media. J. Appl. Phys. **91**, 4659 (2002)

13. K. Svoboda, W. Denk, D. Kleinfeld, D.W. Tank, In vivo dentritic calcium dynamics in neocortical pyramid neurons, Nature **385**, 161 (1997)
14. W. Denk, K. Svoboda, Photon upmanship: why multiphoton imaging is more than a Gimmick. Neuron **18**, 351 (1997)
15. W. Cheong, S.A. Prahl, A.J. Welch, A review of the optical properties of biological tissues. IEEE J. Quantum Electron. **26**, 2166 (1990)
16. S. Morgan, M. Khong, M. Somekh, Effects of polarisation state and scatterer concentration on optical imaging through scattering media, Appl. Opt. **36**, 1560 (1997)
17. A.G. Loewy, P. Siekevitz, *Cell Structure and Function*, 2nd edn. (A Holt International, New York, 1971)
18. W. Ganong, *Review of Medical Physiology*, 16th edn. (Appleton and Lange, Norwalk, 1993)
19. C.F. Bohern, D.R. Huffman, *Absorption and Scattering of Light by Small Particles* (Wiley, New York, 1983)
20. B.R. Masters, P.T.C. So, E. Gratton, Multiphoton excitation fluorescence microscopy and spectroscopy of in vivo human skin. Biophys. J. **72**, 2405 (1997)
21. M. Oheim, E. Beaurepaire, E. Chaigneau, J. Mertz, S. Charpak, Two-photon microscopy in brain tissue: parameters influencing the imaging depth. J. Neuroscience Methods **111**, 29 (2001)
22. K. Svoboda, F. Helmchen, W. Denk, D.W. Tank, Spread of dendritic excitation in layer 2/3 pyramidal neurons in rat barrel cortex in vivo. Nat. Neuronscience **2**, 65 (1999)
23. H.J. Schwarzmaier, A. Yaroslavsky, I. Yaroslavsky, T. Goldbach, T. Kahn, F. Ulrich, P.C. Schulze, R. Schober, Optical properties of native and coagulated human brain structures, in *SPIE Proceedings of Lasers in Surgery: Advanced Characterization, Therapeutics, and System VII* ed by R.R. Anderson, K.E. Bartels, et al. **2970**, 492 (1997)
24. X. Deng, M. Gu, Penetration depth of single-, two-, and three-photon fluorescence microscopic imaging through human cortex structures: Monte Carlo simulation. Appl. Opt. **42**, 3321 (2003)
25. X. Deng, X. Gan, M. Gu, Monte-Carlo simulation of multi-photon fluorescence microscopic imaging through inhomogeneous tissue-like turbid Media. J. Biomed. Opt. **8**, 440 (2003)
26. V. Tuchin, *Tissue Optics* (SPIE Press, Bellingham, 2000)
27. B. Fischl, A.M. Dale, Measuring the thickness of the human cerebral cortex from magnetic resonance images. Proc. Natl. Acad. Sci. **97**, 11044 (2000)

Chapter 9
Image Reconstruction

Abstract The key problem associated with imaging through a turbid tissue medium is the strong scattering effect that severely degrades image quality (Anderson et al., IEEE J Quantum Electron 23:1798–1805, 1987). In order to overcome this problem, approaches based either on the reduction of the scattering effect, e.g., two-photon (2p) fluorescence microscopy (Denk et al., Science 248:73–76, 1990; Saloma et al., Phys Med Biol 43:1741–1759, 1998), or on the use of various optical gating methods, e.g., time gating (Denk et al., Science 248:73–76, 1990; Palmes-Saloma and Saloma, J. Struct. Biol 131:56–66, 2000; Masters and So, Opt. Express 8:2–10, 2001; Fujimoto et al., Opt Lett 3:150–152, 1986), polarization gating (Mackintosh et al., Phys Rev B 40:9342–9345, 1989; Gan et al., Microsc Microanal 3:495–503, 1997; Gan et al., J. Opt. Soc. Am. A 16:2052–2058, 1998), coherent gating (Toida, Appl Opt B 52:391–394, 1991), spatial gating (Fujimoto et al., Opt Lett 3:150–152, 1986), and angle gating methods (Gan et al., Microsc Microanal 3:495–503, 1997), etc., to reduce the number of scattered photons, have been introduced. However, both approaches encounter the same difficulty of retaining strong signal strength. It has been demonstrated that the signal loss due to strong scattered in two-photon (2p) fluorescence microscopy is far more serious than that in single-photon (1p) fluorescence microscopy and that strong optical gating often results in severe signal loss (Gu et al., Appl Phys Lett 77:1551–1553, 2000). In this chapter, we use a novel method for microscopic imaging through turbid media, which allows the collection of scattered photons as well as ballistic and least scattered photons to ensure strong signal strength, and the restoration of image resolution is performed through mathematical image reconstruction.

9.1 Deconvolution

The difficulty of an effective image reconstruction method arises from the fact that mathematical deconvolution requires an accurate knowledge of a point spread function (PSF) [1–11]. If an object is not embedded in a highly scattering medium, a PSF can usually be derived from the setup of an imaging system or be measured

M. Gu et al., *Microscopic Imaging Through Turbid Media*,
Biological and Medical Physics, Biomedical Engineering,
DOI 10.1007/978-3-662-46397-0_9

by the image of a point object. However, if an object is embedded in a highly scattering medium, situation becomes more complicated due to the fact that the surrounding medium plays an important role in image formation. Therefore a PSF without considering surrounding media is an inaccurate measure of image formation and causes considerable distortion in image reconstruction. Our new reconstruction method is based on the concept of the effective point spread function (EPSF), introduced in Chap. 4 [12]. The significance of the EPSF is that it enables us to separate the information about an object from a surrounding turbid medium and an imaging system. Due to the fact that it is impossible to measure the EPSF, the derivation of the EPSF has to rely on mathematical simulation. In this chapter, we utilize the concept of the EPSF and image reconstruction technique to restore the imaging qualities degraded by the multiple scattering process [13].

Based on the concept of the EPSF, the image intensity $I(\boldsymbol{r})$ of a thin object can be modeled by the convolution of an object function $o(\boldsymbol{r})$ and the EPSF $h(\boldsymbol{r})$ (4.1), where $\boldsymbol{r} = (x, y)$ is the transverse coordinate. 10,000,000 illumination photons have been used in the Monte Carlo simulation to ensure the accuracy of an EPSF for a reflection-mode scanning optical microscope. As an example, a turbid medium is assumed to consist of scattering particles (polystyrene beads of diameter 0.5 μm suspended in water, with a concentration of 1 %). We also assume that the excitation and fluorescence wavelengths are, respectively, 488 and 510 nm. According to Mie scattering theory, the corresponding scattering mean free path lengths are 25. 7 and 28.3 μm, and anisotropic values are 0.87 and 0.86, respectively, for excitation and fluorescence wavelengths. The numerical aperture (NA) of the microscope object is assumed to be 0.5 in the simulation, and no confocal pinhole is used to allow the maximum collection of photons.

The simulated images of a thin ring of outer radius 50 μm and inner radius 30 μm at different depths in the turbid medium is illustrated in Fig. 9.1. The reason for choosing such an object is that it emulates biological cells. It is shown that the image becomes blurred with increasing depth into the scattering medium and that the ring structure totally disappears beyond the depth of 150 μm. In order to improve the image resolution, one way is to apply some optical gating techniques to

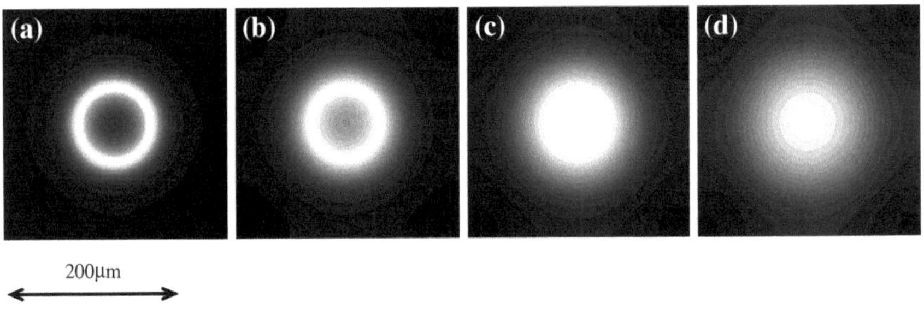

Fig. 9.1 Images of a circular ring of inner radius 35 μm and outer radius 50 μm at different depths in the turbid medium: **a** $d = 100$ μm; **b** $d = 150$ μm; **c** $d = 200$ μm; **d** $d = 250$ μm. Reprinted with permission from [13], 2002, Elsevier

remove scattered photons. However, it would always be a problem in retaining signal strength if a significant amount of scattered photons is removed.

In this chapter, we use the expectation-maximization (EM) algorithm for maximum-likelihood (ML) image restoration [14]. The most basic form of the EM algorithm can be expressed as follows [14]:

$$I^k(\boldsymbol{r}) = h(\boldsymbol{r}) \otimes_2 s^k(\boldsymbol{r}), \tag{9.1}$$

where $I^k(\boldsymbol{r})$ and $s^k(\boldsymbol{r})$ are, respectively, the image intensity and the estimated object function at kth iteration. The predicated image intensity $I^k(\boldsymbol{r})$ is compared with the recorded image intensity:

$$d^k(\boldsymbol{r}) = I(\boldsymbol{r})/I^k(\boldsymbol{r}). \tag{9.2}$$

The ratio $d^k(\boldsymbol{r})$ is then projected back to the object space with a normalization factor $H(0)$:

$$r^k(\boldsymbol{r}) = h(-\boldsymbol{r}) \otimes_2 d^k(\boldsymbol{r})/H(0), \tag{9.3}$$

where $r^k(\boldsymbol{r})$ is a correction factor, and the normalization factor $H(0)$ is derived as the two-dimensional integration of the EPSF. The estimated image intensity of the next iteration can be derived as:

$$I^{k+1}(\boldsymbol{r}) = I^k(\boldsymbol{r}) \times r^k(\boldsymbol{r}). \tag{9.4}$$

The deconvolved images after different numbers of iterations are shown in Fig. 9.2. Here we use the image at d = 200 μm depth as an example. Before the deconvolution process, the image is blurred and ring structure is totally washed out by the blurring effect. It is shown that the image resolution loss due to multiple scattering can be partially recovered through the image reconstruction process. After 200 iterations, the ring structure starts to re-emerge; however, the ring structure is still quite blurred. With more iterations, blurring on the ring structure becomes less significant. It is noted that the images after 1,000 and 2,000 iterations are very similar in terms of sharpness of the image, which indicates that the rate of convergence becomes significantly small and that the image will not be further improved through more iterations.

To further demonstrate the resolvability of the method, we designed an object which consists of two rings. Such an object emulates a biological cell cluster, and the deconvolution process of this object is demonstrated in Fig. 9.3. Before the restoration, the image at d = 200 μm is so blurred that not only the two rings cannot be resolved from each other, but also the individual structure of each ring cannot be identified. After 500 iterations, the two rings can be resolved from each other; however, the structure of each ring cannot be clearly identified. It is also noticed that the spherical symmetry of each ring is poor, which is a common defect occurring in the deconvolution process [15]. When the iteration number increases,

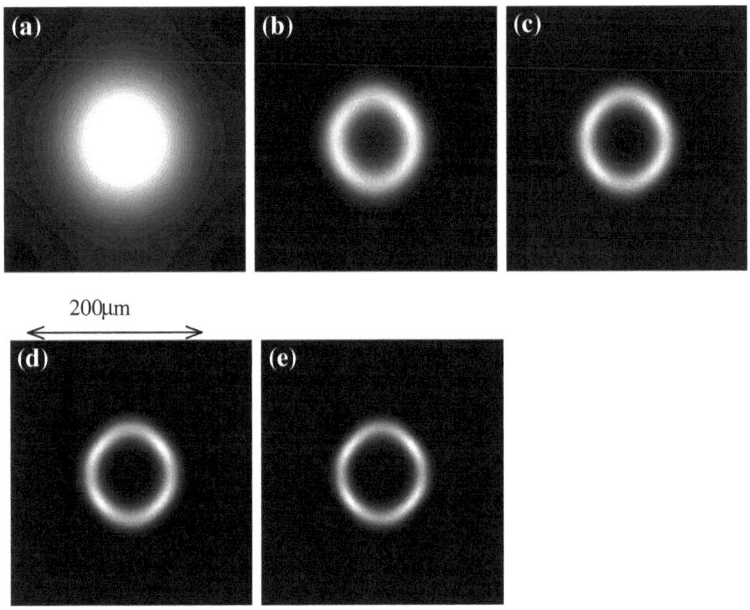

Fig. 9.2 Images of a circular ring: **a** before restoration; **b** after 200 iterations; **c** after 500 iterations; **d** after 1,000 iterations; **e** after 2,000 iterations. Reprinted with permission from [13], 2002, Elsevier

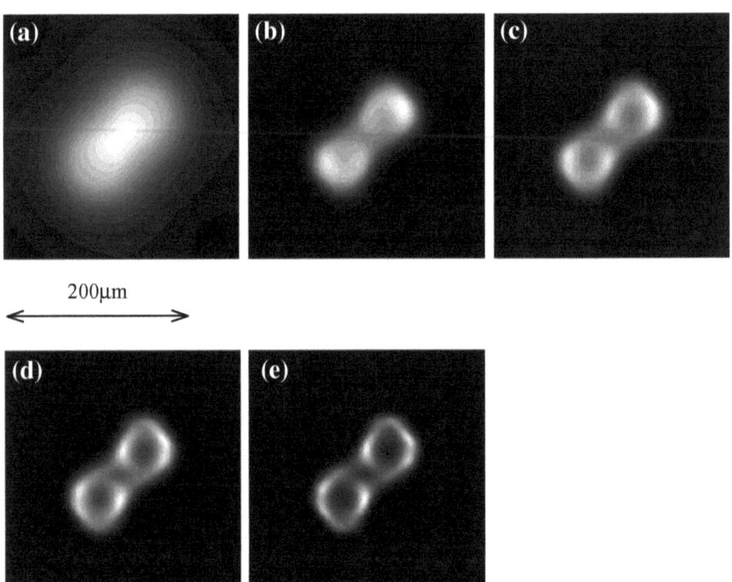

Fig. 9.3 Images of two circular rings: **a** before restoration; **b** after 500 iterations; **c** after 1,000 iterations; **d** after 2,000 iterations; **e** after 4,000 iterations. Reprinted with permission from [13], 2002, Elsevier

the structure of each ring becomes clear and the spherical symmetry is also improved slightly. The comparison between Fig. 9.3d, e shows no significant improvement after 2,000 iterations, which means that numerical stability has been reached.

9.2 Noise Factor

We have demonstrated that with the accurate knowledge of the EPSF, image resolution can be improved significantly by mathematical restoration. However, there is another factor that has not been taken into consideration. Due to the highly scattering nature of tissue-like turbid media, the signal strength is usually weak in a thick sample, which results in poor signal-to-noise ratio. In order to demonstrate such an effect, a random noise is added to the image:

$$I'(\boldsymbol{r}) = I(\boldsymbol{r}) + n(\boldsymbol{r}), \tag{9.5}$$

where $n(\boldsymbol{r})$ is the noise function. In Figs. 9.4 and 9.5, the original images and the restored images of a single ring after 4,000 iterations with random noise of different magnitudes a are illustrated respectively. It is seen that the restored images become distorted if noise is included in the image; the poorer the signal-to-noise ratio, the more severely the image is distorted. Some artifacts also start to appear in the case of strong noise (Fig. 9.5d, e). Therefore, the noise in the image is the main obstacle in image restoration.

Here is the question of how to reduce the noise effect in imaging through a turbid medium. One way to increase signal-to-noise ratio is to increase illumination power. Theoretically it can be effective; however, increasing illumination power always increases the risk of damage to the sample, especially in situations of dealing biological tissues. The other way is to reduce the optical gating effect in an imaging system and collect more scattered photons to serve the purpose of

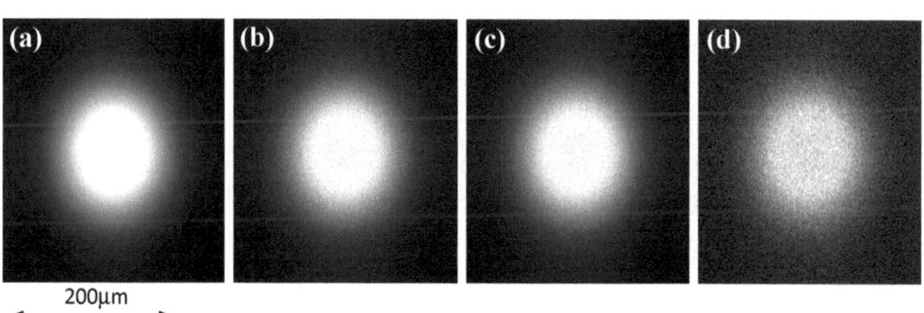

Fig. 9.4 Images of a circular ring of inner radius 35 μm and outer radius 50 μm with noise of different magnitudes a: **a** $a = 0$ %; **b** $a = 10$ %; **c** $a = 20$ %; **d** $a = 30$ %. Reprinted with permission from [13], 2002, Elsevier

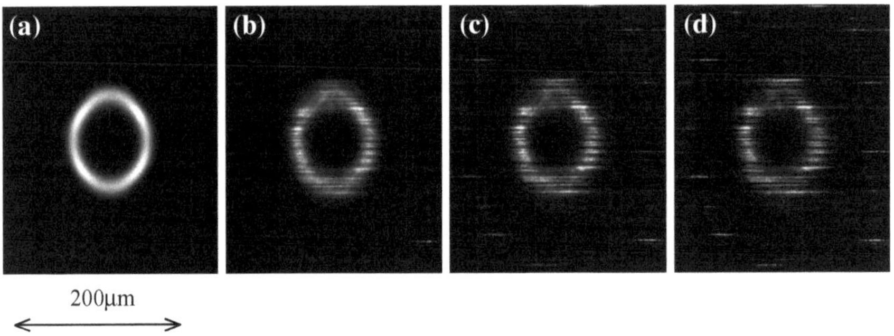

Fig. 9.5 Restored images of a circular ring with noise of different magnitudes a after 4,000 iteration: **a** $a = 0$ %; **b** $a = 10$ %; **c** $a = 20$ %; **d** $a = 30$ %. Reprinted with permission from [13], 2002, Elsevier

increasing signal strength. Although it results in a poor image resolution in this situation, it can be significantly improved through image restoration process with a relatively high signal-to-noise ratio. Optical gating techniques also serve the purpose of improving image resolution, but always result in lower signal level, which brings difficulties in the image restoration process. The results shown in Fig. 9.5 suggest that a sufficient number of scattered photons should be collected to ensure a low noise image and to use the mathematical reconstruction to recover resolution loss.

In conclusion, we have demonstrated a novel method in imaging through a turbid medium. This method is based on the concept of EPSF. With a simulated EPSF, image restoration has been performed based on EM algorithm. The results show that image resolution can be significantly improved through mathematical deconvolution. It is also demonstrated that a high signal-to-noise ratio is crucial in image reconstruction, which can be achieved by collecting scattered photons to improve signal strength.

9.3 Image Reconstruction with Optical Gating

As shown in Chap. 6, polarization gating methods can play an important role in microscopic imaging through turbid media. However, there are some limitations of polarization gating methods, for example, when the degree of polarization $\gamma \rightarrow 0$, the improvement by utilizing polarization gating methods becomes less pronounced. It has also been demonstrated that signal strength can be insufficient if a significant amount of scattered photons is removed. It is worthwhile to discuss the role of scattered photons. Are multiple scattered photons merely noise and make no positive contribution in building an image? The statistical analysis of scattered photon distribution shows that scattered photons still carry information about embedded objects. However, they are always treated as noise when high resolution is pursued. In a thick turbid medium, because of nearly nonexistent ballistic or least scattered photons, multiple scattered photons have to be taken into account in building up an image. This inevitably degrades the image resolution. In this circumstance, the inverse approach (image reconstruction) is regarded as the solution to the problem [16]. Here we use the EM algorithm for ML image restoration [14].

Here we design an object which consists of three rings. Each ring has an outer radius of 20 μm and an inner radius of 15 μm. Such an object emulates a biological cell cluster, and the deconvolution process of this object is demonstrated in Fig. 9.6. We use the image at d = 120 μm depth and without any polarization gating method as an example. The turbid medium consists of 0.48 μm beads. The anisotropy values g is 0.81, for an He–Ne laser at wavelength 0.633 μm. The scattering mean free path length l_s, is assumed to be 20 μm. Before the deconvolution process, the image is blurred and the ring structure is totally washed out by the blurring effect. It is shown that the image resolution loss due to multiple scattering can be partially recovered through the image reconstruction process. After 200 iterations, the ring structure starts to re-emerge; however, the ring structure is still quite blurred and the three ring structures cannot be resolved from each other (Fig. 9.6b). With more iterations, blurring on the ring structure becomes less significant (Fig. 9.6c, d), and there is also a slight indication of resolving three rings after 400 iterations (Fig. 9.6d). It should be pointed out that the EPSF is not only much broader compared with a normal PSF, but also has a significant tail that can affect the image formed. Therefore the cut-off of the EPSF used for deconvolution process needs to be carefully determined.

The deconvolution process of this object under parallel and differential polarization gating methods is shown in Figs. 9.7 and 9.8 respectively. Because some scattered photons have been removed by the optical gating effect, the images before the deconvolution process are better under parallel and differential gating methods, with the best image produced under the differential gating method. Under the parallel polarization gating method, the improvement in image quality is limited, since the amount of scattered photons removed is insignificant. Therefore, the deconvolution process on the image recorded under the parallel polarization gating method has very similar effect compared with that without optical gating methods.

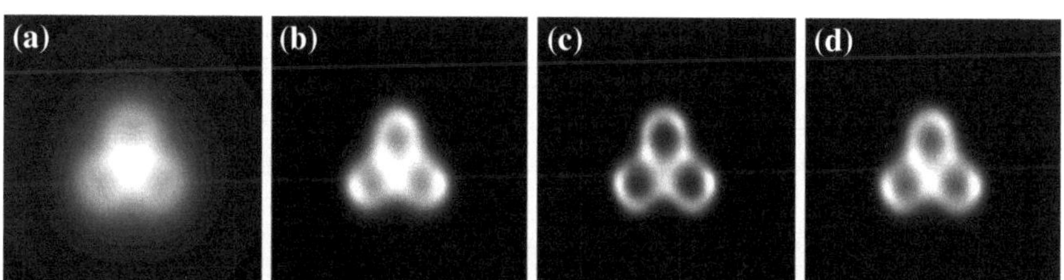

Fig. 9.6 Reconstruction of images without polarization gating methods: **a** before restoration; **b** after 100 iterations; **c** after 200 iterations; **d** after 400 iterations. Reprinted with permission from [16], 2002, SPIE

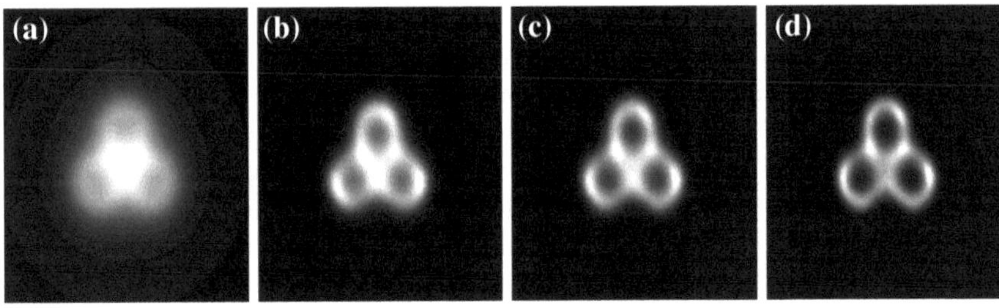

Fig. 9.7 Reconstruction of images under parallel polarization gating methods: **a** before restoration; **b** after 100 iterations; **c** after 200 iterations; **d** after 400 iterations. Reprinted with permission from [16], 2002, SPIE

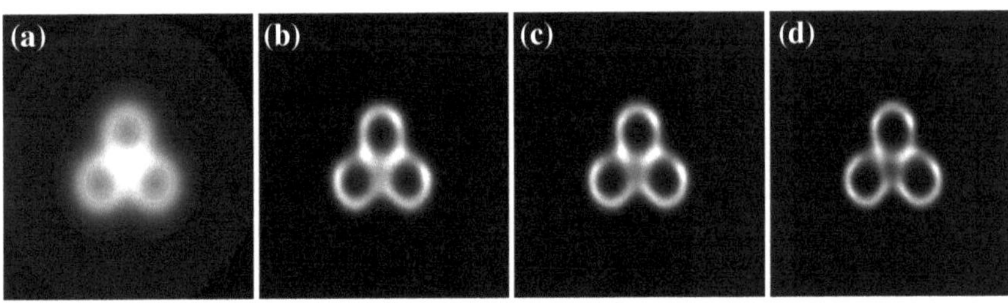

Fig. 9.8 Reconstruction of images under differential gating methods: **a** before restoration; **b** after 100 iterations; **c** after 200 iterations; **d** after 400 iterations. Reprinted with permission from [16], 2002, SPIE

For example, the images after 100, 200, 400 iterations are only slightly better, compared with those respective images recorded without polarization gating methods (Fig. 9.7).

The combination of the differential polarization gating and the image reconstruction methods produces the images with the best quality. It is noticed in Fig. 9.8a that the three ring structures can be vaguely identified even before the deconvolution process, which indicates that the image has been improved significantly through the differential polarization gating method, due to the removal of a large amount of highly scattered photons. After only 100 iterations, each ring structure can be clearly identified (Fig. 9.8b). After 400 iterations, not only the three-ring structure can be resolved, but also the width of each ring becomes thinner and approaches its original width (Fig. 9.8d). This phenomenon shows that with the differential polarization gating method, the deconvolution process can be more efficient.

The comparison of reconstructed images after 1,000 iterations under different gating situations is shown in Fig. 9.9. It is demonstrated that with a stronger gating effect, better reconstructed images can be obtained. It is shown that the width of the rings in Fig. 9.9c is only 6 μm, which is close to the actual width (5 μm) of the object. The comparison of the images in Fig. 9.9 and their respective images after

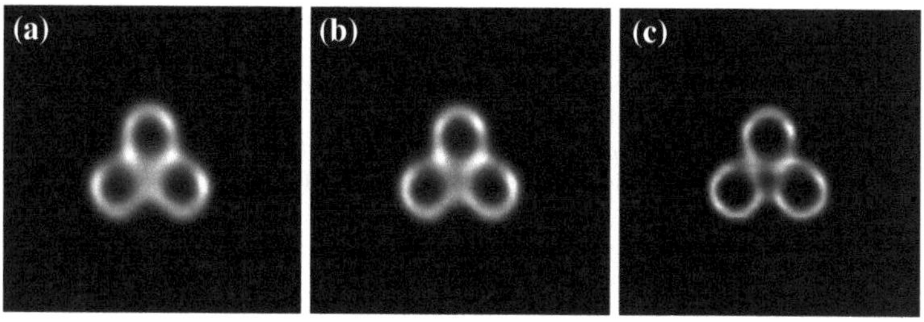

Fig. 9.9 Comparison of reconstructed images after 1,000 iterations: **a** without polarization gating methods; **b** under the parallel polarization method; **c** under the differential polarization gating method. Reprinted with permission from [16], 2002, SPIE

400 iterations shows that the images after 400 and 1,000 iterations are very similar in terms of the width of the rings. This indicates that the rate of convergence becomes significantly small, and that the image will not be further improved through more iterations.

References

1. S. Anderson, S. Montan, S. Svanberg, Multispectral system for medical fluorescence imaging. IEEE J. Quantum Electron. **23**, 1798 (1987)
2. W. Denk, J.H. Strickler, W.W. Webb, Two-photon laser scanning fluorescence microscopy. Science **248**, 73 (1990)
3. C. Saloma, C. Palmes-Saloma, H. Kondoh, Site-specific confocal fluorescence imaging of biological microstructures in a turbid medium. Phys. Med. Biol. **43**, 1741 (1998)
4. C. Palmes-Saloma, C. Saloma, Long-depth imaging of specific gene expressions in whole-mount mouse embryos with single-photon excitation confocal fluorescence microscopy and FISH. J. Struct. Biol. **131**, 56 (2000)
5. B.R. Masters, P.T.C. So, Confocal microscopy and multi-photon excitation microscopy of human skin in vivo. Opt. Express **8**, 2 (2001)
6. J.G. Fujimoto, S. De Silvestri, E.P. Ippen, C.A. Puliafito, R. Margolis, A. Oseroff, Femtosecond optical ranging in biological systems. Opt. Lett. **3**, 150 (1986)
7. F.C. Mackintosh, J.X. Zhu, D.J. Pine, D.A. Weitz, Polarization memory of multiply scattered-light. Phys. Rev. B **40**, 9342 (1989)
8. X. Gan, S. Schilders, M. Gu, Combination of annular aperture and polarization gating methods for efficient in microscopic imaging through a turbid medium: theoretical analysis. Microsc. Microanal. **3**, 495 (1997)
9. X. Gan, S.P. Schilders, M. Gu, Image formation in turbid media under a microscope. J. Opt. Soc. Am. A **16**, 2052 (1998)
10. M. Toida, M. Kondo, T. Ichimura, H. Inaba, Two-dimensional coherent detection imaging in multiple scattering media based on the directional resolution capability of the optical heterodyne method. Appl. Opt. B **52**, 391 (1991)
11. M. Gu, X. Gan, A. Kisteman, M.G. Xu, Comparison of penetration depth between two-photon excitation and single-photon excitation in imaging through turbid tissue media. Appl. Phys. Lett. **77**, 1551 (2000)
12. X. Gan, M. Gu, Effective point-spread function for fast image modeling and processing in microscopic imaging through turbid media. Opt. Lett. **24**, 741 (1999)

13. X. Gan, M. Gu, Microscopic image reconstruction through tissue like turbid media. Opt. Commun. **207**, 149 (2002)
14. T.J. Holmes, Maximum-likelihood image restoration adapted for noncoherent optical imaging. J. Opt. Soc. Am. A **5**, 666 (1988)
15. J.A. Conchello, E.W. Hansen, Enhanced 3-D reconstruction from confocal scanning microscope images. 1: deterministic and maximum likelihood reconstructions. Appl. Opt. **29**, 3795 (1990)
16. X. Gan, M. Gu, Image reconstruction through turbid media under transmission-model microscope. J. Biomed. Opt. **7**, 372 (2002)

Index

M. Gu et al., *Microscopic Imaging Through Turbid Media*,
Biological and Medical Physics, Biomedical Engineering,
DOI 10.1007/978-3-662-46397-0

MIX
Papier aus verantwortungsvollen Quellen
Paper from responsible sources
FSC® C105338

If you have any concerns about our products,
you can contact us on
ProductSafety@springernature.com

In case Publisher is established outside the EU,
the EU authorized representative is:
Springer Nature Customer Service Center GmbH
Europaplatz 3, 69115 Heidelberg, Germany

Printed by Libri Plureos GmbH
in Hamburg, Germany